International Technology Teacher Education

Editor

P. John Williams

55th Yearbook, 2006

Council on Technology Teacher Education

New York, New York   Columbus, Ohio   Chicago, Illinois   Peoria, Illinois   Woodland Hills, California

 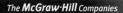

Copyright © 2006 by the Council on Technology Teacher Education.

All rights reserved. Except as permitted under the United States Copyright Act, no part of this publication may be reproduced or distributed in any form or by any means, or stored in a database or retrieval system, without prior written permission of the publisher, Glencoe/McGraw-Hill.

Send all inquiries to:
Glencoe/McGraw-Hill
3008 W. Willow Knolls Drive
Peoria, IL 61614

ISBN 0-07-873461-4

Printed in the United States of America

1 2 3 4 5 6 7 8 9 10 044 10 09 08 07 06

Orders and requests for information about cost and availability of yearbooks should be directed to Glencoe/McGraw-Hill's Order Department, 1-800-334-7344.

Requests to quote portions of yearbooks should be addressed to the Secretary, Council on Technology Teacher Education, in care of Glencoe/McGraw-Hill at the above address, for forwarding to the current Secretary.

This publication is available in microform from

UMI
300 North Zeeb Road
Dept. P.R.
Ann Arbor, MI 48106

FOREWORD

This 55th annual yearbook of the Council on Technology Teacher Education continues the unabridged tradition of scholarly excellence and promotion of discourse in technology teacher education. In this time of globalization and technological change, we are fortunate to have such a forum in these yearbooks to bring to fore the kinds of thinking and innovation that has kept our Council strong and vibrant. This year is no exception. For the second consecutive year we celebrate a first in the history of this yearbook series; we celebrate the assembly of international authors and the first international editor. P. John Williams has assembled a yearbook that could serve as the long needed catalyst for the internationalization of technology education. This yearbook provides our field with critical insight into the varied history, structure, delivery, challenges, and direction of technology teacher education from the perspective of each international author.

Not since section three of the 40th yearbook (1991) of the Council on Technology Teacher Education has even a chapter been devoted to international perspectives on technology education. The contents of this yearbook have been long overdue. Technology education and indeed the world in which we live has changed dramatically as some have characterized, the world as becoming "flatter". Technological developments have enabled citizens of our world to communicate, collaborate, and participate in the economic development of their respective countries. Knowledge and technological capability are quickly becoming the center of the world's economy. Thus, technology education is becoming an international imperative.

The yearbook editor and chapter authors have given careful attention to helping us build an understanding of the diverse approaches to technology teacher education by providing an introduction, history, overview of technology teacher education, structure of technology teacher education, program example, and licensure methods in each of their chapters. The discussion begins with a thorough review of technology teacher education from each of the contributing authors representing 12 different countries. The final chapter helps to synthesize the challenges, commonalities, and bring into focus the varied approaches to technology teacher education.

The editor and chapter authors are to be commended for their insight and treatment of international technology teacher education. We are grateful for their commitment to expanding our perspectives and provoking thought and conversation about the internationalization of technology teacher education. We have much in common and will grow even closer as in the words of author Thomas L. Friedman, The World is Flat.

On behalf of the Council and the Yearbook Committee, we are honored to present this yearbook to the profession. The Council is grateful to have Glencoe McGraw-Hill publishing company as our partner in the yearbook series. Their shared commitment to technology teacher education has made a significant contribution to the field and is truly appreciated. Finally, I join with the Council membership in once again thanking everyone who has contributed to these remarkable series of scholarly works since 1952.

> Michael A. De Miranda
> President, CTTE
> March 2006

YEARBOOK PLANNING COMMITTEE

Terms Expiring 2006
    Patrick Foster
        Central Connecticut State University
    Edward M. Reeves
        Utah State University

Terms Expiring 2007
    Rodney L. Custer, Chairperson
        Illinois State University
    Michael A. De Miranda
        Colorado State University
    G. Eugene Martin
        Texas State University, San Marcos

Terms Expiring 2008
    Kurt R. Helgeson
        St. Cloud State University
    Linda Rae Markert
        State University of New York at Oswego

Terms Expiring 2009
    Roger B. Hill
        The University of Georgia
    Doug Wagner
        Manatee School District, Bradenton, Florida

Terms Expiring 2010
    Mark Sanders
        Virginia Polytechnic Institute & State University
    William L. Havice
        Clemson University

OFFICERS OF THE COUNCIL

President
Michael A. De Miranda
Colorado State University
School of Education
Fort Collins, CO 80523

Vice-President
Richard D. Seymour
Ball State University
Department of Industry & Technology
Muncie, IN 47306

Secretary
Michael K. Daugherty
Technology Education
University of Arkansas
107 Graduate Education
Fayetteville, AR 72701

Treasurer
Marie C. Hoepfl
Appalachian State University
Department of Technology
Boone, NC 28608

Past President
Rodney L. Custer
Illinois State University
Department of Technology
Normal, IL 61790-5100

YEARBOOK PROPOSALS

Each year at the ITEA International Conference, the CTTE Yearbook Committee reviews the progress of yearbooks in preparation and evaluates proposals for additional yearbooks. Any member is welcome to submit a yearbook proposal, which should be written in sufficient detail for the committee to be able to understand the proposed substance and format. Fifteen copies of the proposal should be sent to the committee chairperson by February 1 of the year in which the conference is held. Below are the criteria employed by the committee in making yearbook selections.

CTTE Yearbook Committee

CTTE Yearbook Guidelines

A. Purpose
   The CTTE Yearbook Series is intended as a vehicle for communicating major topics or issues related to technology teacher education in a structured, formal series that does not duplicate commercial textbook publishing activities.

B. Yearbook topic selection criteria
   An appropriate yearbook topic should:
   1. Make a direct contribution to the understanding and improvement of technology teacher education;
   2. Add to the accumulated body of knowledge of technology teacher education and to the field of technology education;
   3. Not duplicate publishing activities of other professional groups;
   4. Provide a balanced view of the theme and not promote a single individual's or institution's philosophy or practices;
   5. Actively seek to upgrade and modernize professional practice in technology teacher education; and,
   6. Lend itself to team authorship as opposed to single authorship.

   Proper yearbook themes related to technology teacher education may also be structured to:
   1. Discuss and critique points of view that have gained a degree of acceptance by the profession;
   2. Raise controversial questions in an effort to obtain a national hearing; and,
   3. Consider and evaluate a variety of seemingly conflicting trends and statements emanating from several sources.

C. The Yearbook Proposal
   1. The yearbook proposal should provide adequate detail for the Yearbook Committee to evaluate its merits.
   2. The yearbook proposal includes the following elements:
      a) Defines and describes the topic of the yearbook;
      b) Identifies the theme and describes the rationale for the theme;
      c) Identifies the need for the yearbook and the potential audience or audiences;
      d) Explains how the yearbook will advance the technology teacher education profession and technology education in general;
      e) Diagram symbolically the intent of the yearbook;
      f) Provides an outline of the yearbook which includes:
         i) A table of contents;
         ii) A brief description of the content or purpose of each chapter;
         iii) At least a three level outline for each chapter;
         iv) Identification of chapter authors (s) and backup authors;
         v) An estimated number of pages for each yearbook chapter; and,
         vi) An estimated number of pages for the yearbook (not to exceed 250 pages).
      g) Provides a timeline for completing the yearbook.

   It is understood that each author of a yearbook chapter will sign a CTTE Editor/Author Agreement and comply with the Agreement. Additional information on yearbook proposals is found on the CTTE web site at http://teched.vt.edu/ctte/.

PREVIOUSLY PUBLISHED YEARBOOKS

*1. *Inventory Analysis of Industrial Arts Teacher Education Facilities, Personnel and Programs,* 1952.
*2. *Who's Who in Industrial Arts Teacher Education,* 1953.
*3. *Some Components of Current Leadership: Techniques of Selection and Guidance of Graduate Students; An Analysis of Textbook Emphases;* 1954, three studies.
*4. *Superior Practices in Industrial Arts Teacher Education,* 1955.
*5. *Problems and Issues in Industrial Arts Teacher Education,* 1956.
*6. *A Sourcebook of Reading in Education for Use in Industrial Arts and Industrial Arts Teacher Education,* 1957.
*7. *The Accreditation of Industrial Arts Teacher Education,* 1958.
*8. *Planning Industrial Arts Facilities,* 1959. Ralph K. Nair, ed.
*9. *Research in Industrial Arts Education,* 1960. Raymond Van Tassel, ed.
*10. *Graduate Study in Industrial Arts,* 1961. R.P. Norman and R.C. Bohn, eds.
*11. *Essentials of Preservice Preparation,* 1962. Donald G. Lux, ed.
*12. *Action and Thought in Industrial Arts Education,* 1963. E.A.T.Svendsen, ed.
*13. *Classroom Research in Industrial Arts,* 1964. Charles B. Porter, ed.
*14. *Approaches and Procedures in Industrial Arts,* 1965. G.S. Wall, ed.
*15. *Status of Research in Industrial Arts,* 1966. John D. Rowlett, ed.
*16. *Evaluation Guidelines for Contemporary Industrial Arts Programs,* 1967. Lloyd P. Nelson and William T. Sargent, eds.
*17. *A Historical Perspective of Industry,* 1968, Joseph F. Luetkemeyer Jr., ed.
*18. *Industrial Technology Education,* 1969. C. Thomas Dean and N.A. Hauer, eds.; *Who's Who in Industrial Arts Teacher Education,* 1969. John M. Pollock and Charles A. Bunten, eds.
*19. *Industrial Arts for Disadvataged Youth,* 1970. Ralph O. Gallington, ed.
*20. *Components of Teacher Education,* 1971. W.E. Ray and J. Streichler, eds.
*21. *Industrial Arts for the Early Adolescent,* 1972. Daniel J. Householder, ed.
*22. *Industrial Arts in Senior High Schools,* 1973. Rutherford E. Lockette, ed.
*23. *Industrial Arts for the Elementary School,* 1974. Robert G. Thrower and Robert D. Weber, eds.
*24. *A Guide to the Planning of Industrial Arts Facilities,* 1975. D.E. Moon, ed.
*25. *Future Alternatives for Industrial Arts,* 1976. Lee H. Smalley, ed.
*26. *Competency-Based Industrial Arts Teacher Education,* 1977. Jack C. Brueckman and Stanley E. Brooks, eds.
*27. *Industrial Arts in the Open Access Curriculum,* 1978. L.D. Anderson, ed.
*28. *Industrial Arts Education: Retrospect, Prospect,* 1979. G. Eugene Martin, ed.
*29. *Technology and Society: Interfaces with Industrial Arts,* 1980. Herbert A. Anderson and M. James Benson, eds.
*30. *An Interpretive History of Industrial Arts,* 1981. Richard Barella and Thomas Wright, eds.

*31. *The Contributions of Industrial Arts to Selected Areas of Education,* 1982. Donald Maley and Kendall N. Starkweather, eds.
*32. *The Dynamics of Creative Leadership for Industrial Arts Education,* 1983. Robert E. Wenig and John I. Mathews, eds.
*33. *Affective Learning in Industrial Arts,* 1984. Gerald L. Jennings, ed.
*34. *Perceptual and Psychomotor Learning in Industrial Arts Education,* 1985. John M. Shemick, ed.
*35. *Implementing Technology Education,* 1986. Ronald E. Jones and John R. Wright, eds.
 36. *Conducting Technical Research,* 1987. Everett N. Israel and R. Thomas Wright, eds.
*37. *Instructional Strategies for Technology Education,* 1988. William H. Kemp and Anthony E. Schwaller, eds.
*38. *Technology Student Organizations,* 1989. M. Roger Betts and Arvid W. Van Dyke, eds.
*39. *Communication in Technology Education,* 1990. Jane A. Liedtke, ed.
*40. *Technological Literacy,* 1991. Michael J. Dyrenfurth and Michael R. Kozak, eds.
 41. *Transportation in Technology Education,* 1992. John R. Wright and Stanley Komacek, eds.
*42. *Manufacturing in Technology Education,* 1993. Richard D. Seymour and Ray L. Shackelford, eds.
*43. *Construction in Technology Education,* 1994. Jack W. Wescott and Richard M. Henak, eds.
*44. *Foundations of Technology Education,* 1995. G. Eugene Martin, ed.
*45. *Technology and the Quality of Life,* 1996. Rodney L. Custer and A. Emerson Wiens, eds.
 46. *Elementary School Technology Education,* 1997. James J. Kirkwood and Patrick N. Foster, eds.
 47. *Diversity in Technology Education,* 1998. Betty L. Rider, ed.
 48. *Advancing Professionalism in Technology Education,* 1999. Anthony F. Gilberti and David L. Rouch, eds.
*49. *Technology Education for the 21st Century: A Collection of Essays,* 2000. G. Eugene Martin, ed.
 50. *Appropriate Technology for Sustainable Living,* 2001, Robert C. Wicklein.
 51. *Standards for Technological Literacy: The Role of Teacher Education,* 2002. John M. Ritz, William E. Dugger, and Everett N. Israel, eds.
 52. *Selecting Instructional Strategies for Technology Education,* 2003. Kurt R. Helgeson and Anthony E. Schwaller, eds.
 53. *Ethics for Citizenship in a Technological World,* 2004. Roger B. Hill, ed.
 54. *Distance and Distributed Learning Environments: Perspectives and Strategies,* 2005. William L. Havice and Pamela A. Havice, eds.

*Out-of-print yearbooks can be obtained in microfilm and in Xerox copies. For information on price and delivery, write to UMI, 300 North Zeeb Road, Dept. P.R., Ann Arbor, Michigan 48106.

PREFACE

There is at least a dual rationale for this book. One relates to the increasing internationalization of the International Technology Teacher Association and consequently its Council on Technology Teacher Education. To wit I am proud to be the first non-US editor of a CTTE yearbook, and I am sure I will not be the last.

The world of technology education continues to become smaller as international collaboration takes place through research, curriculum development and conferences. This parallels technology itself, as no technology is independent, and the technology of no country is isolated.

The second rationale is the state of technology teacher education, and the only absolute statement that can be made about its status is that it is dynamic. In an increasingly regulated and competitive environment, teacher education programs are responding to a range of pressures while at the same time struggling to be authentic and contemporary. As this book describes the issues and responses in 12 countries, it is hoped that the sharing of information and ideas will help maintain a constructive momentum in technology teacher education.

A range of criteria were used to select the countries and authors for this book, related to author expertise, internationalization and the technology education system. The final selection of countries provides a broad range of contexts and the authors are all renowned experts.

55th Yearbook Editor
P. John Williams

ACKNOWLEDGMENTS

I would like to acknowledge Glencoe/McGraw Hill for its continued support of Technology Education through the council's Yearbook Series. The series has become a valuable professional resource for technology educators around the world.

I would also like to thank the members and officers of the Council on Technology Teacher Education for their initial reception of my proposal for this book, and for the feedback and guidance provided at the Yearbook meetings. I thank them for tolerating my absence when not always able to travel to the US for meetings, for John Ritz's representations at those times, and for his guidance during the formative stages of the book. I would also like to acknowledge the expert proofreading of Anita Kreffl.

Finally, I would like to primarily acknowledge the authors of the chapters of this yearbook. Their belief in the importance of Technology Teacher Education in a global environment stimulated them to see the project through what at times seemed like interminable drafts and reviews to its conclusion. They were chosen because they were professional friends, and I am pleased to be able to say that we remain friends after the completion of this book. They are also eminent educators in the context of their countries, and I trust that the publication of this book will enhance their eminence. They are:

Frank Banks is Director of the Centre for Research and Development in Teacher Education (CReTE) in the Faculty of Education and Language Studies at the Open University (OU) in the United Kingdom and was until recently director of the innovative flexible teacher-training programme. This programme won the Queen's Award for Higher Education in 1994. Frank teaches from Undergraduate to Doctoral level and most recently was an author of technology modules for the new on-line teacher professional development website *TeachandLearn.net,* a collaboration between the OU and the BBC. He has been a Visiting Professor in the School of Engineering and Advanced Technology, Staffordshire University (2000-2003), a sub-dean for the faculty and has acted as a consultant to the World Bank and to Egyptian, South African and Argentinean government agencies. Frank has authored or edited twelve books and handbooks for teachers, including the well known *Teaching Technology,* and over forty academic papers in the field of teacher professional knowledge and teacher education, particularly using open and distance learning techniques.

Moshe Barak received his D.Sc. and M.Sc. degrees in Science and Technology Education from the Technion, Israel Institute of Technology, were he served as a lecturer and researcher from 1986 to 2002. Dr. Barak has conducted several research projects for curriculum development, teacher enhancement and educational evaluation, and has served as a Program Referee at the Ministry of Education Centre for Science and Technology Teaching, next to the Weitzman Institute of Science. He is a member of the Editorial Board of the International Journal of Technology and Design Education. Currently, Dr. Barak is a Senior Lecturer at the Graduate Program for Science and Technology Education at Ben-Gurion University of the Negev in Israel. His research interests focus on curriculum development, teacher training and promoting pupils' learning skills and creative thinking.

Jacques Ginestié is married with two children and is Professor and deputy director of the Institut Universitaire de Formation des Maîtres d'Aix-Marseille, France, in charge of secondary teacher training. He is Head of the investigation team Gestepro, part of the Unité Mixte de Recherche Apprentissage, Didactique, Évaluation, Formation. This team develops investigation in science education, technology education and vocational training education. He is also Coordinator of the Réseau Inter Universitaire de Formation Initiale et Continue d'Enseignant d'Education Technologique (RIUFICEET) and President of the scientific committee of the Réseau Africain des Institutions de Formation de Formateurs de l'Enseignement Technique (RAIFFET). His personal field of investigation is the study of the teaching-learning process in technology education through observation of the school organisation through the interactions between subject, teacher and pupil. The main methodology developed is based on the task and activity analysis.

Gerd Höpken is Akademischer Oberrat (associate professor) at Flensburg University, Germany. Since 1968 he has been involved in the transition from arts and crafts to technology education as a school teacher, research assistant, and lecturer. As an ITEA "Ambassador" he is the connecting link between technology education in Germany and other countries. Together with a colleague he translated the Standards of the "Technology for all Americans" project (Vol I and II) as well as the book "Technically Speaking" into German. His current research interest is comparing technology education in different countries as well as the methodology of technology and technology education. He has published more than 60 articles (mainly in German and English, but also in French and Spanish) and authored and co-authored 10 books and edited and co-edited 4 books. He regularly presents at international and national conferences.

Juan L. Iglesias has been Faculty Dean of Humanities and Education at the University of Atacama, Chile since 1987. He was Associate Professor of Educational Research at South Chile University and is currently Chair and Professor at the University of Atacama. He is the creator and current chairman of the *Teachers for the 21st Century* which is a new curriculum model based on and inspired by the Problem Based Learning concept. He has been working in initial teacher education for more than forty years in Normal Schools and Universities in Chile. Since 1998 he has been Vice-President of the National Council of Chilean Education Faculties. He was also appointed to several national commissions for the improvement of the Chilean Education System. In 2002 he was appointed a member of the National Ministry of Education Advising Council. In 2003 he was selected as a member of the National Commission for the Chilean System of Educational Assessment Reform (SIMCE), and in 2004 he was appointed to the National Commission for Research on distance education programs for Initial Teacher Training in Chile.

Alister Jones is currently Director of the Wilf Malcolm Institute of Educational Research, School of Education at the University of Waikato, New Zealand and Chairperson of the Board of Trustees of Technology Education New Zealand. He is the former director of the Centre for Science and Technology Education Research. He has been involved in research in science and technology education in both England and New Zealand. He has been extensively involved in technology education in New Zealand since 1992 and has been director of a number of technology education contracts, including policy advice to the Ministry of Education, technology curriculum development, national facilitator training and teacher development resource, and research into student and teacher learning and assessment. In 2000 he was awarded the New Zealand Science and Technology Medal for contributions to technology education.

Hidetoshi Miyakawa is an internationally recognized researcher and teacher educator of the Japanese Society of Technology Education and an honoured member of Epsilon Pi Tau. He was presented with the Prakken Professional Cooperation Award in 2001 and was a recipient of the Distinguished Technology Educator award in 2003. In over 15 years as a university professor, he has hosted international researchers and collaborated on research in technology education in the United States and abroad. Hidetoshi's research interest is in how technology education contributes to the enhancement of creativity, self-education, and problem solving among students. He has designed a unique assessment instrument to monitor and confirm student development and progress across these areas, and his research is designed to assist in the development of effective teaching materials and curricula to extend and enhance these areas.

Chris Mothupi is married to Ntswaki and has three children (Boipelo, Natefo and Othupile). He is a director of PROTEC, a large Mathematics, Science and Technology Education organisation in South Africa that provides in-service training and classroom based support to teachers. Initially a science teacher, he started to develop an interest in Technology Education when the subject was introduced in South African schools in the early nineties. He was a member of the Ministerial Project Committee that revised the national curriculum where he chaired a committee of experts that developed the Revised National Curriculum Statement for Technology for Grades R–9. He is a co-author of three Technology Education school textbooks. His research interests include learner assessment and project evaluation.

Margarita Pavlova is the Director of Technology Education in the Faculty of Education at Griffith University in Australia. She is leader of the project on Education for Sustainable Development in the Centre for Learning Research at Griffith, and Adjunct Professor at Nizhny Novgorod Institute of Development of Education, Russia. She has a PhD from the Russian Academy of Education, and another from La Trobe University, Australia. She has been Scientific Director of the "Technology & Enterprise Education in Russia" program for the past ten years. The results of that program are being implemented at the National level through a tender won with the World Bank for writing new sets of Technology education texts for Y5–Y9. She has been the recipient of a number of research and professional development grants. Her research interests include: the conceptualisation of technology education; comparative education; and education for sustainable development. She is an author and co-author of 7 books, 5 textbooks, and over 17 book chapters and articles.

Mark Sanders is Professor and Program Leader, Technology Education, Virginia Polytechnic Institute and State University in the USA. He earned a Ph.D. from the University of Maryland (1980) and has taught Technology Education at the high school and university levels since 1975. His research interests include the implementation of communication technologies in educational settings; the integration of technology education with science, engineering, and mathematics curricula; and the status of technology education in the US. Since 1989, he has served as founding Editor and Associate Editor of the *Journal of Technology Education*. National offices held include Vice-President and Treasurer of the Council on Technology Teacher Education and President and Vice-President of the International Graphic Arts Education Association. In 1992, he established the *e-JTE* as one of the first academic e-journals in history and in 1997 established GRAPHIC COMM CENTRAL, then-and still-the Web portal for graphic communication teachers and students. His many publications include a popular communication technology textbook, 11 CTTE Yearbook chapters, and an array of monographs and journal articles.

Andrew Stevens is a lecturer in Technology Education at Rhodes University, Grahamstown, Eastern Cape, South Africa. He also co-ordinates the in-service Advanced Certificate in Education programme for the Faculty of Education. Prior to this he was a branch manager of the ORT-STEP Institute, a pioneering NGO in the field of technology education in South Africa. He has spent most of his professional life in and around schools, having started as a High school mathematics teacher and ended his school career as the principal of one of the first non-racial schools in South Africa. His research interests include problem-solving in mathematics and technology education and development of the field of technology teacher education in South Africa.

Arley Tamir studied in Electronics (B.Sc.) and in Science Teaching (M.Sc.) before receiving a Ph.D. degree at the Hebrew University in Jerusalem. He is considered to be one of the founders of Israel's Technology Education System. Arley has served as: National Supervisor for Technology Education, Chairman of The Public Committee for Setting Israel's Technology Education Objectives Towards the Year 2000, Chairman of the 2nd Jerusalem International Science and Technology Education Conference (JISTEC '96), Deputy-Director of The Science and Technology Education Department in the Ministry of Education, Culture and Sport, Project-Manager of The Joint Thai-Israeli Project for the Establishment of the National Curriculum and Teaching Aids Center in Bangkok, Thailand and as Director for Research and Development at The Holon Technology Education Center, University of Tel-Aviv. In 1997 he was elected as Head of ORT Jerusalem Academic College for Technology Teachers. Arley is presently engaged in many national and international activities aimed to promote Technology Education.

Kenneth S. Volk is Principal Lecturer at The Hong Kong Institute of Education. Ken received his Ph.D. from the University of Maryland and was Assistant Professor at East Carolina University before joining the Institute in 1995. During his ten-years at HKIEd, Ken has been instrumental in designing the first Bachelor of Education program and well as Design and Technology facilities at the new campus. He teaches technical and professional subjects in Design & Technology education, and was recognized by the Institute in 2004 for his distinguished teaching. His research has focused on gender issues, which has led to opportunities opening up for all students in Hong Kong. Ken continues to be an active member of professional organizations such as ITEA and CTTE, as well as serving on the editorial board of several journals.

P. John Williams lectures in the School of Education at Edith Cowan University in Perth, Western Australia, and coordinates undergraduate, postgraduate and offshore programs in technology education. Apart from Australia, he has worked and studied in a number of African countries and in the United States. He has recently directed the nationally funded Investigation into the Status of Technology Education in Australian Schools. His current research interests include design collaboration at a distance, and distance and on-line technology education. He has published over 40 articles and has authored and co-authored 8 books, regularly presents at international and national conferences, consults on Technology Education in a number of countries, is a longstanding member of 8 professional associations and is on the editorial board of 4 professional journals.

55th Yearbook Editor
P. John Williams

CONTENTS

Foreword iii
Yearbook Planning Committee v
Officers of the Council vi
Yearbook Proposals vii
Previously Published Yearbooks viii
Preface x
Acknowledgments xi

Chapter 1: Technology Teacher Education in Australia 1
    P. John Williams
    Edith Cowan University, Australia

Chapter 2: Technology Teacher Education in Chile 23
    Juan. L. Iglesias
    Universidad de Atacama, Chile

Chapter 3: Technology Teacher Education in France 45
    Jacques Ginestié
    Professeur Des Universités, France

Chapter 4: Technology Teacher Education in Germany 69
    Gerd Höpken
    Flensburg University, Germany

Chapter 5: Technology Teacher Education in Hong Kong 89
    Kenneth S. Volk
    The Hong Kong Institute of Education, Hong Kong

Chapter 6: Technology Teacher Education in Israel 111
    Moshe Barak
    Ben-Gurion University of the Negev, Israel
    Arley Tamir
    Science & Technology Education Systems Consultant, Israel

Chapter 7: Technology Teacher Education in Japan 129
Hidetoshi Miyakawa
Aichi University of Education, Japan

Chapter 8: Technology Teacher Education in New Zealand 147
Alister Jones
University of Waikato, New Zealand

Chapter 9: Technology Teacher Education in Russia 167
Margarita Pavlova
Griffith University, Australia

Chapter 10: Technology Teacher Education in South Africa 189
Chris Mothupi
PROTEC, South Africa
Andrew Stevens
Rhodes University, South Africa

Chapter 11: Technology Teacher Education in the United Kingdom 215
Frank Banks
Open University, UK

Chapter 12: Technology Teacher Education in the United States 241
Mark Sanders
Virginia Polytechnic Institute & State University, USA

Chapter 13: Technology Teacher Education: Summary 271
P. John Williams
Edith Cowan University, Australia

Technology Teacher Education in Australia

Chapter I

P. John Williams
Edith Cowan University, Australia

INTRODUCTION

The public education system in Australia is provided by each of the five state and two territory governments, so specifics of school starting age, the curriculum and the division of primary and secondary schooling, vary across Australia. Students usually begin school at ages 5–6, with primary school spanning 6–7 years and secondary school a further 5–6 years. School is compulsory to age 15, so generally the last two years of secondary schooling are post-compulsory. There is an increasing emphasis on retaining students at school throughout the post compulsory years.

The state and territory governments also support non-government schools, which enrolled about 28% of primary and 37% of secondary students in 2002 (MCEETYA). Most government schools are co-educational, but a significant number of non-government schools are single-gender schools. The federal government provides some funding to all schools to support specific priorities and strategies, but the majority of government school funding comes from state and territory governments and part of non-government school funding comes from the federal government.

An increasing number of schools fall outside the traditional primary-secondary categories. For example middle schools cover upper primary and lower secondary grades, and some secondary schools are designated Technology High Schools, Technology Colleges or Technical Colleges. A development in 2005 was the establishment of Technical Colleges by the federal government. These colleges will cover education for students in the final two years of secondary schooling and have an emphasis on vocational education. This direct funding of schools represents a new role for the federal government.

In 1994, *A Statement on Technology for Australian Schools* (Australian Education Council) was published, as a result of all the states and territories cooperating on developing frameworks in eight learning areas, one of which was technology. Since then all the states and territories have established technology learning areas through the development of curriculum, support material and professional development. Various titles have been adopted in different states, including Technology Education, Technological and Applied Studies, Technology and Enterprise, but they contain similar elements. There is a significant degree of consistency in the definitions of technology used by education systems in Australia. Technology is defined broadly, and key common elements of the definitions include 'the application of knowledge and resources' used 'to extend human capabilities'. There is strong general agreement that technology involves a process, that is, there is an identifiable method used in the development of technology. This process is most commonly referred to as design, but it is not defined or described in detail. Similarly the relationship between the concepts or knowledge of technology and the processes of technology is not explored extensively.

There are few curricula in technology that describe an accompanying body of knowledge, though in some instances new subjects have been developed with the introduction of technology as a learning area. This has left teachers to modify existing subjects to conform to the new approach.

All curricular descriptions of technology in Australia address the importance of relationships between technology, society and the environment. From the descriptions given in curriculum documents, technology is presented as something people develop and use, with some recognition of the reverse relationship that technology influences and shapes human behaviour.

Technology Education in Primary Schools

The incorporation of technology education in the primary years of schooling is a relatively new phenomenon in Australia and is not yet universal. Despite this, 90% of primary schools indicated that they taught technology in all grades (Williams, 2001). The emphasis of much of this technology was Information Technology, embedded in a range of subjects, rather than relating to broad technology education. School-based decision making and curriculum planning mean that there is a variety of technology education occurring at a classroom level. Often, the curriculum programs in this area are determined by individual interests and enthusiasm of the teachers and principals, and the educational priorities in particular schools.

The amount of time designated to the teaching of technology is determined at school level, and differences are evident. In some states there is currently no specific time allocated to the teaching of technology in primary schools. Only one state has adopted a specific syllabus: Science and Technology. Schools in all other states devise their syllabus and classroom activities from either state or national frameworks.

In most states and territories the amount of student participation in technology related learning varies from a 'theoretical' one eighth of the curriculum (being one of eight learning areas) or four hours per week to less than one hour per week. However, it is likely that there is less technology being taught in primary schools now than 10 years ago when the learning area began, and this is probably because of the recent focus on literacy and numeracy (Williams, 2001).

In 1997 the Australian Science and Technology Education Council (ASTEC) reported on the teaching of science and technology in Australian primary schools. They found an increased level of interest in technology education by teachers, evidence that technology education can facilitate the general education of children, and a recognition of the potential for science and technology education to be brought together in a complementary way.

ASTEC also made a number of findings that they considered to be of concern, relating to the development of quality technology education. In terms of professional development, ASTEC noted that some good programs had been developed and it was critical these professional development programs be sustained. The Prime Minister's Science, Engineering and Innovation Council (1999) also found that many primary teachers had outdated or insufficient technology teaching qualifications.

In primary schools, technology education is generally delivered through an integrated approach with other learning areas, though there are some discrete technology programs. Many primary teachers are still coming to grips with the notion of technology education generally, and together with inadequate experience and training, they lack confidence and competence in the content, as well as the practice; for example, using basic hand and power tools with resistant materials and electronics.

The relationship between technology education generally and the area of Information Technology and computer studies is not clear, and the terminology tends to be inter-changeable. In many primary schools there is a focus on computers, but not on other areas of technology education.

Technology Teacher Education in Australia

Technology Education in Secondary Schools

Technology Education is delivered through a range of technology related subjects in the secondary school including Home Economics, Technical Studies, Computing, Information Technology, Media, Industrial Arts, Design and Technology, Agriculture and Business Studies. In all states and territories, technology education (as delivered through these subject contexts) is either a centrally mandated part of the junior secondary curriculum, or the majority of schools ensure students study some technology.

For those schools offering technology education in Years 7 and 8 it is most common to offer up to four introductory subjects. At Years 9 and 10 a range of technology subjects are offered, commonly as electives and in Years 11 and 12 a narrower range of technology education subjects is offered. These elective subjects may work together, for example through a school organized technology learning area, to achieve the outcomes for technology education. Otherwise, they may operate as independent subjects as a choice within the school timetable and so compete with each other for students. Such decisions are school based, and the trend is for individual technology subjects to be integrated more as a learning area.

For example, in the state of New South Wales all students are expected to complete 200 hours of Design and Technology in Years 7–8 including an integrated component of 50 hours computing-related study. In Tasmania the technology-related-subjects in years 7–10 require 100 hours per year. In Years 7–8 technology studies are offered in Materials, Design and Technology, Food and Textiles and Information Technology. In Western Australia all schools need to show that students have demonstrated the Technology and Enterprise outcomes from the curriculum in all compulsory years of schooling.

In no state or territory are technology subjects compulsory in Years 11–12. The significant trend is that students are able to take an increasing number of technology subjects and use the examination scores from those subjects for tertiary entrance, an option that has not been widely available in the past.

The difficulties faced in the implementation of technology education in primary and secondary levels of schooling are different. Primary teachers lack knowledge and expertise to give them the confidence in technology, and secondary schools have a tradition of providing independent technical subjects from which technology education must develop. Progression for students is difficult to map, and the links between these two levels of schooling are not strong.

HISTORY

The first real efforts to introduce technology education into the curriculum of the state schools occurred towards the end of the nineteenth century. The 1880's saw Australia in the depths of a depression. This economic situation existed until the mid 1890's when economic recovery took place with significant growth. The impact of this situation saw the need for: "...restructuring the economy in a way that focused attention on the growing need for a higher level of technical and commercial skills in the workforce" (Laird, 1982, 74).

Murray-Smith (1966, 418), although referring to technical education in general, made a very apt statement concerning the influence of the 1880's and 1890's depression upon the public's perception of the role of technology education: the viewpoint that in slow economic times it was part of the duty of technical education to take over some of the training responsibility of the individual employer. This sentiment, though referring to the period of the late eighteenth Century, describes what was to become a pattern in technology education in Australia. This pattern is that in times of depressed economic activity, new levels of awareness of the role of technology education are the result.

Later curricula were based on the United Kingdom models of education, primarily those from Scotland and Wales (Williams, 1993), dividing practical education into commercial, technical and domestic areas at the post primary level. Primary graduates of the time were directed to the schools most appropriate to their perceived academic aptitudes, and students who were identified as being more academically well equipped were directed to the more traditional classic curriculum. Students who were less academically capable were directed to the gender specific technical, domestic or commercial areas.

The intention for these schools was that they train artisans through a vocational-based curriculum, including subjects that were drawn from the common trades of the time: woodwork, metalwork, trade drawing in the technical schools, and cooking, hygiene and sewing in the domestic schools. However, these technical studies schools never reached the popularity that was expected of them.

Technical education was to receive little attention during the 1930's and 1940's, because it was during these years two significant periods of social upheaval occurred: the Great Depression and World War II. The 1950's was a time of fast economic growth in the industrial sector, providing an over abundance of employment opportunities, so there was little attention on the technology component of the curriculum. The attention it did receive resulted in technical subjects being broader than required for the specific trades and were more aligned with general rather than vocational education.

By the late 1980's, technology education was to again become an area of interest. This period was characterised by poor economic performance and there were again calls for an increase in the profile of technology education within the school curriculum.

In 1987, the Australian Education Council (AEC) began a series of initiatives that led to the publication in 1994 of nationally agreed curriculum statements and profiles related to eight learning areas, one of which was technology. This declaration of technology as a learning area had profound implications. Firstly, all subject areas in secondary schooling from which technology education developed were located within the elective areas of the curriculum. The implication was that these subjects provided specific learning experiences relevant only for specific groups of students with particular interests or career destinations in mind. Indeed, some of these subjects were regarded by students and the community as being relevant only to a particular gender. Secondly, in the case of primary education, technology had not generally been part of school programs, and primary teachers had little experience to draw on when developing programs.

A *Statement on Technology for Australian Schools* (Australian Education Council, 1994) set out what was regarded as the technology learning area. This included the place of technology in society, the need for all students to experience technology education and the form in which it should appear in the school curriculum. It outlined four strands for learning in technology education: Designing, Making and Appraising; Information; Materials; and Systems. These were regarded as interdependent and were intended to be developed sequentially through stages or levels in the compulsory years of schooling.

Prior to the 1990's, school curricula addressed technology in a very limited way. In the main, technology was referred to in elective or optional syllabuses. Most often students' perceptions of technology were developed from a very restricted range of learning experiences, for example, students might learn about the tools and machines used to work with timber. Invariably, learning focused on an established body of technical 'know-how'. In some courses students learnt about designs that characterised past eras.

At the secondary school level, technology education has tended to develop out of vocational studies, but in the past decade studies have been introduced to courses that address 'technology' in a more systematic and comprehensive way, and are grounded in a more general than vocational philosophy of education. These courses integrate the use of technology processes and encourage students to make value judgments and to be creative and innovative. From entry to year 12, students are required to develop projects, practice management skills and engage in independent and group learning. These studies aim to develop students' qualities of flexibility, adaptability and enterprise.

Probably the most significant aspect of the change to technology education is the concept that as a learning area it contributes to all students' general education and therefore should be studied by all students in the compulsory years of schooling. The breadth and dynamic nature of technology itself is necessarily reflected in this technology education. This is a positive educational attribute resulting in a healthy diversity of approaches across Australia to the teaching and study of technology. At the same time, this diversity provides challenges related to national curriculum development and teacher support.

The status of technology in the curriculum is variable across the states and systems in Australia. In some states for example, technology subjects are compulsory and in others they are elective, though they are offered in some form in 95 percent of the schools (Williams, 2002). Technology education continues to evolve and change. As the states and territories develop their own curriculum, the level of diversity across Australia increases. The curricula are moving away from the structures and content outlined in *A Statement on Technology for Australian Schools* (Australian Education Council, 1994) which had the initial effect of a very similar technology curriculum across all the states.

OVERVIEW OF TECHNOLOGY TEACHER EDUCATION

Around 1990, primary and secondary teachers for Australian schools were trained at Colleges of Advanced Education. Around this time the process of amalgamating these institutions began, some colleges combining and becoming new universities, while others merged with existing universities.

The rationale for the training of technology education teachers to become part of universities' responsibilities when the amalgamation took place was that those in the university technological faculties could strengthen the content base of technology teacher training. The integration of teacher training into universities has only succeeded to a very limited extent. "There is little evidence that those in the technological faculties recognise a responsibility to play a service role in teacher training" (Watts, 1998, p.10).

In 1996, a survey of technology teacher education programs in Australia (Williams) indicated that of the 38 universities in Australia, nine were identified as offering undergraduate technology teacher education programs. All these institutions offered a four year Bachelor of Education degree in technology education. The degree was variously appended as Design and Technology, Technological and Applied Studies, Technology Education, or Technology. All the training programs were under the control of the institutions' Faculty of Education.

Since this survey was conducted in 1996, many institutions have moved away from this model, for example to a double degree structure or to a graduate entry pre-service training model. An additional development has been the demise of secondary undergraduate technology teacher education courses in four states, and an increase in post-graduate technology courses for both secondary and primary teacher trainees in all states except one. There are an increasing range of entry and exit points in training programs, developing links between universities and technical training institutions and a number of courses which have been designed for specific client groups.

For a number of reasons, technology teacher training programs are problematic for universities. Gibson and Barlow (2000) outline the problems facing universities when they endeavour to provide a range of degree programs. Universities allocate Equivalent Full Time Student Units (EFTSUs) to their various programs based on the institution's academic profile, which is negotiated annually with the federal government. This in turn determines the total Australian government funding that each university receives. Over a number of years this source of funding has decreased, and as a result, universities are having to rely more on non-government income sources, including financial endowments from private persons and industry, commercial enterprises, as well as an increasing willingness to enrol full fee paying students (local and international) to alleviate the budgetary constraints. Gibson and Barlow claim that technology teacher education programs are not attractive to university administrators: they have suffered from low intake of students, they are perceived to be expensive, they provide limited opportunities for economies of scale through large lectures and they are less likely to attract fee-paying students than other courses.

In a report to the Australian Academy of Technological Sciences and Engineering (ATSE), Watts (1998) elaborated on the problems in technology teacher training at universities, "The training of teachers is now a university responsibility. It is one in which they have a long history of a lack of real commitment. The intellectual culture of universities leads to attitudes in which they accept willingly a role in training only for the influential professions...not the practice of teaching... Teachers in training clearly do not benefit from the excellence of the faculties and the professional competence of the academic staff in science and engineering" (p.35).

A consideration for the introduction of post-graduate pre-service courses is for first degree students to become suitably qualified, in that they needed to have undertaken an initial degree relevant to technology education. This remains an issue for all states offering teacher training at the post-graduate levels, largely because of the breadth of technology offered in schools and the relative specialization of most initial degrees.

Four year undergraduate programs offer a range of specializations including design and technology, food technology, home economics, textile technology, engineering science, computing studies, business and technics. The study of content, curriculum and education is concurrent throughout all four years.

A more recent approach to training technology teachers, stimulated by the shortage of these teachers, is the design of programs to suit specific groups of clients, sponsored by the state governments through fees and other payments, as a way of ensuring an adequate supply of technology teachers. These courses vary from one to two years in length and include significant components of school-based practice.

Another increasingly common trend is for university technology teacher training programs to make links with technical colleges for the provision of the skills-based machine and production oriented aspects of the course. The logic is that the technical colleges have specialized machinery for the training of apprentices and trades people, and as it is expensive to update and maintain equipment, this represents a sensible consolidation. A philosophical difference arises however as the technical college approach is competency based and generally teacher-oriented, contrasted to the technology education approach which is generally outcomes-based and student-centered through a designerly approach to making. Some universities have rejected this liaison after trialing it for a number of years and others continue with a significant technical college component to their course.

The number of students studying vocational courses at senior secondary school has increased dramatically in the past ten years. Many of these vocational areas are closely aligned with technology education and are taught by technology teachers. Some states have implemented additional requirements for teachers to teach in these vocational areas, for example a qualification from a technical college. This has resulted in additional pressure for universities to develop links with technical colleges in order to satisfy these additional vocational teaching qualifications, and to some extent the vocationalization of the technology teacher education program. However, quality in technology teacher education is dependent on a research-based, practical study of a range of industries and technologies and a critical approach to the social and environmental contexts of technology, not a study of a narrow range of specific vocations.

There are still some primary teacher training programs at universities in Australia which do not provide any instruction in technology education, despite the establishment of technology as one of the core learning areas since 1989. However, the majority of training programs offer at least one compulsory unit in technology education, and a number allow for specialization in the technology area by taking additional elective units. These specialized graduates often become technology resource teachers in a school or a region.

Because technology education in schools has developed from a number of previously independent subjects, teachers' professional associations are still organized along the lines of these subjects; for example, at the national level, organizations supporting technology teachers are the Home Economics Institute of Australia, Australian Council for Education through Technology, Design in Education Council Australia, Australian Council for Computers in Education, Council of Australasian Media

Education Organizations, Business Educators of Australia and National Association of Agricultural Educators. These organizations have traditionally supported secondary teachers, and while some are developing support for primary teachers, this is not their main focus. The organizations in the states and territories reflect this national structure; there is no professional organizational support for both primary and secondary teachers, which covers the breadth of the technology learning area. This deficiency of a broad national teacher's professional association and advocacy body is reflected in significantly less federal government attention to technology education than other school learning areas.

Teacher Shortage

The shortage of technology teachers in Australia occurs in a context of low appeal of teaching as a career. "Those with technological competence recognise better career prospects elsewhere. Graduates entering schools have available only limited term contracts and no clear career prospects. These realities compound and contribute to teacher discontent and a lack of public support" (Watts, 1998, p13).

Although the future supply of teachers was predicted to be sufficient to satisfy overall teacher demand in the short and medium term, when each of the states and territories were considered, there were consistent shortages of teachers in technology areas (MCEETYA, 1998; Preston, 2001). A number of other authors (Gibson and Barlow, 2000; Yon, 1999) have indicated a shortage of technology teachers in different states.

While there is a current shortage of technology teachers available to meet this continuing demand, a number of state government employers have responded by actively investing in strategies aimed at alleviating the problem through specially developed Diplomas of Education, targeted retraining programs, student sponsorships, promotion of technology teaching in schools, and recruiting from other states and overseas. In counterbalance to this last initiative, other countries are targeting Australia as a source of technology teachers.

STRUCTURE OF TECHNOLOGY TEACHER EDUCATION

There is no national control over the structure or the content of teacher education courses in Australia. It is entirely up to the university, and in some states, the state department of education, as the major employer of graduates, accredits courses proposed by the universities. Consequently there is a significant degree of diversity in the structure of technology teacher education courses across the states and territories in Australia.

In a survey conducted in 1996 (Williams), all the technology teacher education courses in the country had the same structure, which were four year Bachelor of Education BEd) programs. In 2004, very few four year BEd programs remain. There are a number of possible reasons for this change. One is the national trend in all areas of teacher education for double degrees. Another is the trend to design courses for a specifically targeted group of clients.

A further reason is the desire for faculties of education to cease the provision of what is perceived to be expensive workshop laboratories and equipment necessary for the practical aspects of technology teacher education. In a context where government funding to universities has been reduced by about 28% over the past eight years, there is pressure to reduce the costs of education. An additional impetus to outsource workshop activities is because workshop groups are necessarily small and cannot be reorganized into mass lectures to develop economies of scale. An indication of this trend is the rise in university student:staff ratios from 12.9:1 to 18.8:1 in the ten years since 1990. The specific outcomes of these changes will be discussed in more detail below.

Primary Technology Teacher Education

A recent review of teacher education programs (Chester, 2002) indicated that 33 of the 39 universities in Australia offer teacher education courses for primary teachers. Of the 33 universities, nine do not stipulate units in technology as a compulsory part of their course program, with the majority offering one unit in a 32 unit course, as a compulsory core studies unit in technology.

The interpretation of technology is however quite broad. Two of the courses were Information Technology only, 13 were designated Science and Technology, one Science, Technology and Society, one Technology and the Arts, one Science Technology and Numeracy, and eleven were Technology. So it seems that about one third of primary teacher education courses have at least one compulsory unit devoted specifically to technology, and another third have at least part of a unit devoted to technology (Chester, 2002). In addition, the structure of some courses permits students to take more technology units as electives and so develop further expertise in this area.

These courses cover a broad range of approaches to technology; including a focus on the process, integration with Information and Communication Technologies, the content of technology (materials, structures, mechanisms and electronics for eg) and a minority focusing on using tools and equipment.

Secondary Technology Teacher Education

Because the learning area of technology is quite broad in schools, students training to be secondary teachers generally specialize from one to a number of areas. Depending on the state or territory within Australia, the technology learning area may include the subjects of design and technology, materials design, engineering, home economics, food technology, textiles and design, industrial arts, agriculture, business and computing. Very few universities offer all specializations, but rather limit the options available to students.

The area of training secondary technology teachers is where significant diversity of structure exists between institutions. Some examples follow:

- **Four-year Bachelor of Education.** This is the traditional model and a number of institutions still offer this course as the mode of training technology teachers. Education, pedagogy and content units are studied concurrently throughout the four years. School practice experiences begin in the first or second year, and generally conclude with an extensive period (for example, one term of 10 weeks) working in a school assuming the full teaching load of a teacher.

    In this model the majority of the units in the course are taught by the faculty of education, although some specialized areas may be contracted out to other departments within the university or to external institutions such as secondary schools or technical colleges. In some universities, all practical work is done at either a secondary school or technical college. The first year may be a year at a technical college, and students with this requisite may gain admission into year two at university.

    Applicants may receive advanced standing into the BEd if they have relevant experience or qualifications in a relevant area such as a technical college diploma or a trade apprenticeship. Students may also gain entry to this degree program through completion of an Associate Degree which has less demanding entry qualifications.

- **Two-year Bachelor of Teaching.** This is an integrated post-graduate course offered to students from varying backgrounds. Applicants must have an initial three year under-graduate degree in a technology related area such as engineering, industrial design, architecture or computing. Students focus on education, curriculum and pedagogical studies with the assumption that their content expertise is derived from their initial degree, although there is the option to undertake study in technology subject areas other than those covered in their first degree.

  As alternative entry to the Bachelor of Teaching, a one-year bridging course is available for technical college graduates who would typically have a trade background. Variations on this structure include a two year Bachelor of Education, which follows a similar structure to the two-year Bachelor of Teaching, and a three year Bachelor of Teaching.

- **One-year Graduate Diploma of Education.** The prerequisites for entry into the Graduate Diploma are similar to the Bachelor of Teaching: a three year initial degree in a technology related area. Students then focus on education, curriculum and pedagogical studies, and spend a significant proportion of course time in schools developing pedagogical skills and school knowledge.

- **Two-year Graduate Diploma.** Entry into this award can be through a number of means: certified substantial relevant work experience, including an apprenticeship of not less than eight years experience, a two year technical college qualification plus work experience. The prerequisite qualifications and experience again provide the area of content expertise, so the course units focus on education, pedagogy and curriculum studies.

- **Four-year double degree.** This award is either two degrees undertaken concurrently, or a 2+2 structure where one degree is completed before the other begins, but either pattern is completed within four years. One degree is typically offered by the faculty of education, and the other by the area in which the students specialize. The education degree is generally a Bachelor of Education or a Bachelor of Teaching, and examples of the other degree include a Bachelor of Technology, Bachelor of Design, and Bachelor of Design and Technology.

Unlike other curriculum areas such as science or mathematics, technology has no identifiable disciplinary home within a university. Students consequently study their content areas from departments such as engineering, graphic and industrial design, business, hospitality, multimedia/ computing and art. This is one of the rationales for the double degree; that it enables teacher trainees to develop their content knowledge from a disciplinary approach rather than from a curriculum or school education approach.

Apart from their accreditation role, state departments of education have had an impact on teacher training programs through sponsorship of students in specific courses, and the design of 'tailor-made' courses in order to address the shortage of technology teachers. Examples of such courses include a one year course for three-year trained primary teachers, an 18 month course for retrenched industry workers which includes a six month school internship, a modified three semester Graduate Diploma for targeted applicants, six months of education followed by six months of school mentoring for secondary teachers who studied in other specializations.

AN EXAMPLE: EDITH COWAN UNIVERSITY

All the secondary teacher education programs at Edith Cowan University have changed to a double degree structure over the past five years. The technology program has been the last to make this change, partly because of the difficulty of identifying appropriate content areas from within other faculties of the university.

The Bachelor of Arts (Education) is an existing award that can only be taken in conjunction with another degree in the double degree format, in this case a Bachelor of Design (Technology) that is awarded by the School of Communications and Multimedia. This joint award is limited to those who complete the combined degree. If students wish to exit earlier without completing the Education award, they are able to do so through the Bachelor of Design after three years of study. There are no set advanced standing regulations for applicants, but students with relevant experience or qualifications may apply for unit exemptions.

The design of the degree was based on a contemporary philosophy linking trends in workplace demands, government priorities and school curriculum developments. Society requires creative individuals who are able to communicate well, think originally and critically, adapt to change, work cooperatively, remain motivated when faced with difficult circumstances and are capable of finding solutions to problems as they occur – in short individuals with the array of skills constituting a well developed capacity for innovation. These skills are outlined in *Backing Australia's Ability* (Commonwealth of Australia, 2001) – the innovation action plan devised by the Federal government with the goal of promoting innovation in Australia. Emphasis in this course is on developing Design and Technology teachers with an increased emphasis on creativity and design, and is in line with the priorities of Australian governments.

The Bachelor of Arts (Education) / Bachelor of Design (Technology) degree will prepare students to teach Technology, which has as its goal the development of cognitive skills such as critiquing, analysing, solving problems, generating innovative ideas and communicating what they do, as well as manipulative and organizational skills through making their designs. These objectives are addressed in each Design and Technology unit within the course.

This double degree equips secondary teachers to teach design and technology in the compulsory years of secondary schooling where the approach is both broad and general and covers a range of materials and processes. The graduates will be the first cohort specifically prepared to teach the new post-compulsory courses of study that are being introduced in Western Australian schools from 2006. The new courses related to Technology include Engineering Studies; Materials Design and Technology, and Visual Communications and Design. Table 1 outlines the structure and content, by unit, of this course.

Table 1. Bachelor of Arts (Education) / Bachelor of Design (Technology) degree

Year 1
Semester One

Unit Code	Unit Title	Credit Points
EDS1101	Issues, Challenges and Directions in Education	15
VIS1108	Foundation 1 – Level 1	15
IMM1120	Introduction to Digital Technologies and Multimedia	15
DTM1104	Materials Design and Technology 1	15

Semester Two

Unit Code	Unit Title	Credit Points
EDS1201	Learning, Curriculum and ICT in Schools	15
VIS1208	Studio Practice 1 (Studio Foundation) – Level 2	15
IMM1121	Digital Photomedia	15
DTM1105	Materials Design and Technology 2	15

Year 2
Semester One

Unit Code	Unit Title	Credit Points
EDS2101	Principles and Practices of Teaching	15
DTE2171	Design and Technology Education 1	12
DTC2101	Visual Communication and Design 1	15
DTF2101	Foundations of Design	15

Semester Two

Unit Code	Unit Title	Credit Points
EDS2102	Classroom Management and Motivation	15
VIS2309	Studio Practice 6 – Level 4	15
TEE2117	Technology and Enterprise Education 1	10
DTM2104	Materials Design and Technology 3	15
PPA2172	Professional Practice 1	4
PPA2272	Professional Practice 2	4

Year 3
Semester One

Unit Code	Unit Title	Credit Points
DTS3101	Systems Design 1	12
DTS3102	Systems Design 2	15
DEN3101	Engineering Design 1	15
DTC3101	Visual Communication and Design 2	15
PPA3376	Professional Practice 3	3

Semester Two

Unit Code	Unit Title	Credit Points
EDS3103	Diversity and Inclusivity in the Secondary Classroom	15
EDS4157	Literacy Across the Curriculum	12
DTE3271	Design and Technology Education 2	10
EN3102	Engineering Design 2	10
PPA3477	Fourth Professional Practice	10

Year 4
Semester One

Unit Code	Unit Title	Credit Points
EDS4131	Post-Compulsory Education for Lifelong Learning	10
EDS4130	Middle Years Education	10
DTE4371	Design And Technology Education 3	10
PPA4571	Assistant Teacher Program	30

Semester Two

Unit Code	Unit Title	Credit Points
	School of Education elective	15
DEN4103	Engineering Design 3	15
DTP4101	Design Project	15
DTP4102	Curriculum Development Project in D&T	15

This course is a concurrent form of teacher education where the student is able to combine studies in Education and Design and Technology in each semester. In the early stages, greater attention is given to content studies while professional studies tend to dominate the latter semesters. School practice experiences begin in year two and culminate in the Assistant Teacher Program (ATP) of ten weeks in the final year.

In the content units, theory and practice are integrated, so a three hour block of contact time cannot be divided into one hour of theory and two hours of practice. Teaching and demonstration time is flexible, with much teaching being on-demand. A recently redesigned technology building facilitates this arrangement with workshop areas organized and wired for effective teaching.

Students are often engaged in large projects which last for an extended period of time. They are taught and work under lecturer supervision in the laboratory or workshop for a prescribed number of hours each week, and are expected to work for at least the same number of hours at other times during the week. The workshops are open for about ten hours each day, and the computer labs have 24 hour access.

TEACHER CERTIFICATION

As the Australian states and territories are educationally independent, each has a separate system for teacher certification (registration), though reciprocal recognition exists for teachers who move from state to state. The state which most recently implemented a system of teacher registration was Western Australia in 2004. This state will be used as an example in this instance. Prior to 2004 teachers were deemed certified if they possessed the appropriate teaching qualification granted by a university. An Act of Parliament in 2004 established the Western Australian College of Teaching (WACOT), one role of which is to enforce teacher registration.

A number of levels of registration have been implemented. The requirements for full registration are that the applicant:

(a) holds a qualification in teaching approved by the College for registration (an approved teaching qualification has at least four years of tertiary education, including at least one year of teacher education and demonstrates a minimum of 45 days successful teaching practice);

(b) has not been convicted of an offence which renders the person unfit to be a teacher;

(c) has successfully completed a prescribed police criminal record check;
(d) has achieved standards of professional practice approved by the College;
(e) is proficient in the English language both written and oral;
(f) within the 5 years preceding the application has been teaching for at least one year.

Registration as a teacher expires after five years and may be renewed.

A category of Provisional Registration has been developed for teachers that have been out of the profession for at least five years. During the provisional membership period, the College will provide applicants with guidelines that outline the process to become fully registered. The emphasis is on collaborative work, professional reflection and mentor support. The requirements for provisional registration are that the applicant:

(a) holds a qualification in teaching approved by the College for provisional registration;
(b) has not been convicted of an offence which renders the person unfit to be a teacher;
(c) is proficient in the English language both written and oral.

Provisional registration as a teacher expires after three years and may be renewed.

An additional category is Limited Authority to Teach (LAT) that enables people who do not meet the qualifications' standard to be employed in schools to undertake teaching roles. Schools that are unable to fill vacancies with registered teachers may be allowed under the LAT system to employ people who are not qualified as teachers for up to two years. The requirements for LAT are that the applicant:

(a) has specialist knowledge, training, skills or qualifications;
(b) is proficient in the English language both written and oral;
(c) has been offered a teaching position for which a suitable teacher is not available;
(d) has not been convicted of an offence which renders the person unfit to be a teacher.

CONCLUSION

Technology Education is well established as one of the eight learning areas. In primary schools it is still developing as more teachers become familiar with the area, and in secondary schools it is developing from a technical tradition. Because of this development, and the fact that the states and territories are educationally independent, there is significant diversity of practice in schools. In at least one state, as the curriculum is restructured, there are indications that in future, Technology Education may no longer form part of the core of studies.

There is no single national professional association that represents all technology teachers. This inhibits development of the profession through the production of support material, the fostering of links and advocacy with governments and other external bodies. Australia is comparatively under-populated, and the existence of seven different educational systems across five states and two territories, leads to duplicity of effort and no economies of scale.

The diversity of practice in schools is mirrored by a diversity in both structure and content of technology teacher education. There is little research comparing different models of technology teacher education, and new course development seems to be a response to pressures other than 'designing the best course', such as finances or shortages. Trends in course design include multiple exit and entry points, various/varying length of courses depending on prior learning, and links with schools and technical colleges for the teaching of workshop areas. There is a shortage of technology teachers, a situation that will continue for some years.

REFERENCES

Australian Education Council. (1994). *A statement on technology for Australian schools.* Carlton: Curriculum Corporation.

Australian Science and Technology Education Council. (1996). *Matching Science and Technology to Future Needs 2010.* Canberra: ASTEC.

Australian Science, Technology and Engineering Council. (1997). *Foundations for Australia's Future: Science and Technology in Primary Schools.* Canberra: AGPS.

Commonwealth of Australia (2001) *Backing Australia's Ability.* Canberra: Commonwealth of Australia.

Chester, I. (2002). Australian Technology Teacher Education Programs: their Structure and Marketing. *Initiatives in Technology Education – Comparative Perspectives: Technical Foundation of America Forum.* Gold Coast, Australia, January.

Gibson, J., & Barlow, J. (2000). *NSW Technology Teacher Education: Y2K a time for optimism?* Paper presented at the Biennial National Conference of the Australian Council for Education through Technology, Canberra, ACT.

Ministerial Council on Education, Employment, Training and Youth Affairs. (1998). *Draft School Teacher Demand and Supply Primary and Secondary.* Carlton South: MCEETYA.

Ministerial Council on Education, Employment, Training and Youth Affairs. (2000). *New Pathways for Learning.* Canberra: MCEETYA.

Murray-Smith, S. (1966). *A History of Technical Education in Australia,* PhD. Thesis, Melbourne University.

Preston, B. (2000). *Teacher supply and demand to 2005: projections and context.* A report commissioned by the Australia Council of Deans of Education: Canberra.

Prime Minister's Science, Engineering and Innovation Council. (1999c). *Ideas for innovation.* Canberra: Department of Industry, Science and Tourism.

Prime Minister's Science, Engineering and Innovation Council. (1999b). *Raising awareness of the importance of science and technology to Australia's future.* Canberra: Department of Industry, Science and Tourism.

Prime Minister's Science, Engineering and Innovation Council. (1999a). *Strengthening the Nexus between Science and its Applications.* Canberra: Department of Industry, Science and Tourism.

Ramsey, G. (2000, February/March). Graduate Teachers are Towards the low end of the Experience Scale. *Education Review, The Australian,* p. 13.

Watts, D. (1998). *A Report on the Readiness of Australian Schools to Meet the Demands of Teaching the Curriculum Areas of Science and Technology in the Compulsory Years of Schooling.* Perth, WA: University of Notre Dame, Australia.

Williams, A. (1996). An Introduction to Technology Education, Chapter 1 in Williams, P.J., & Williams, A. (Eds.) *Technology Education for Teachers.* Melbourne: Macmillan

Williams, A.P. (1993). *The Rational for Technical Education in New South Wales Secondary Schools,* M.Curr.St., Dissertation, University of New England

Williams, P.J. (1996). Survey of Undergraduate Secondary Technology Teacher Training programs in Australia. *Australian Journal of Research in Technology and Design Education,* 4(1), 11–15.

Williams, J., & Williams, A. (Eds.). (1996). *Technology Education for Teachers.* Melbourne: MacMillan Education Australia Pty Ltd.

Williams, P.J. (2000). Technology Education in Australia: Where to Now? *1st Biennial International Conference on Technology Education Research.* GoldCoast, Australia, December.

Williams, P.J. (2001). *The Teaching and Learning of Technology in Australian Primary and Secondary Schools.* Department of Education, Science and Technology Working Report, Commonwealth of Australia.

Williams, P.J. (2002). Crisis in Technology Education in Australia. *Second Biennial International Conference on Technology Education Research.* Gold Coast, Australia, December.

Williams, P.J. (2003). The Status of Technology Education in Australia. *Initiatives in Technology Education – Comparative Perspectives: Technical Foundation of America Forum.* Gold Coast, Australia, January. http://teched.vt.edu/ctte/HTML/Research1.html

Yon, R. (1999). Critical shortage of Technology Studies and Home Economic Teachers. *Technology News,* 15 (2), 4–6.

Technology Teacher Education in Chile

Chapter 2

Juan. L. Iglesias
Universidad de Atacama, Chile

INTRODUCTION

The Chilean educational system (K–12) is ninety percent financed by the Central Government and is mainly controlled by the Ministry of Education and its administrative dependencies. The Ministry of Education acts as a coordinator and is in charge of developing, evaluating and supervising all aspects of Chilean education. In addition, the Ministry defines and determines general education policies and practices and maintains programs for improving the quality and ensuring equity in the general education system (Ministry of Education, Chile 2000).

In Chile, pre-school education includes children up to five or six years; primary or elementary school, levels/grade 1–8, are for children between six and 15 years; and secondary school, levels/grade 9–12 are for students from 16 to 21 years. Elementary and secondary school education is compulsory for all Chilean citizens. There is also a tertiary, or higher, educational level composed of three types of private and state education institutions: 1) technical training centres which are supervised by the Minister of Education, 2) professional institutes which are supervised by the National Higher Education Council (NHEC) and 3) the private and state universities, 16 of which are autonomous by law. Several of the private universities obtained autonomy as a result of a decision by the NHEC.

From 1995 the Chilean general education system has been the subject of much debate and discussion, the results of which have been curriculum reform throughout the entire educational system (K–12). The central goal of this reform is to redefine and restructure the educational objectives and contents of study programs across the curriculum (K–12). This is being done in an attempt to help students meet the challenges and needs of a continually evolving society and help them develop the capacities necessary to understand and apply the new knowledge that society continually produces (Ministry of Education, Chile. 2002).

One of the positive results of this new reform is that it is now possible to broaden technology education in Chile to include the concept of technology literacy, which has been defined by the International Technology Education

Association as "the ability to use, manage, assess and understand technology" (ITEA 2000, p.9). It should be mentioned that technology education, as discussed throughout this chapter, also includes two other traditional areas of study in the technological field: 1) vocational education, which refers to the technical training students receive in the final two years of secondary technical school, and 2) technical education, which refers to the technical training that students receive at tertiary institutions such as technical training centres, professional institutes and universities.

Problems and Progress in Chilean Technology Education

The Chilean government, after intensive and extensive research and consultation, has come to the realization that there are three serious problems within the education system: 1) that the primary and secondary curricula do not include technology education, 2) that study programs leading to certification of technicians are currently inefficient and obsolete (this becomes especially problematic when the excessively large number of technician classifications is considered) and 3) the vocational schools, the technical training centres and the professional institutes are currently using tools, machinery, equipment and pedagogical applications that are both obsolete and ineffective in providing the technical awareness, skills and expertise to satisfactorily assist in the industrial development of the country.

Fortunately, with respect to the new reform (1999) it is now possible to include within the category of technology education, technological literacy from levels 1–12. In addition, at the end of 2001, the Ministry of Education approved the new curriculum changes, and the initial objectives of technological literacy are now included in kindergarten. The new reform has also eliminated traditional manual arts courses, such at wood/ metal work and elementary horticulture, which were common in the system during the past few decades.

Briefly, the principal objectives of the new curriculum in technology literacy in the first cycle of primary education (levels 1–4) are currently: 1) assisting the student in the discovery, exploration and understanding of the physical environment, 2) recognition of the various modes of human intervention and interaction that take place, and can take place, within that environment, and 3) assisting the student to understand the many diverse types of activities that develop within this environment and its relevant professions (Elton, 2002).

At grade/levels 5–6 in primary education, the study programs are now structured in three units: in level 5 the sequence is as follows: 1) the historical evolution and social impact of a technological object, 2) analysis of an object and 3) the development and maintenance of an object. At grade/level 6 the sequence is: 1) analysis of technological objects as services, 2) technological systems as an association of components and 3) repair of simple objects.

The study program at level 7 introduces the students to subjects associated with the changes that human technology produces in the environment when modifying and using natural resources. At level 8 the program offers activities for the analysis and understanding of subjects related to the use of technology in different productive processes, such as, the knowledge that objects are composed of physical systems and subsystems that make their operation possible, the use of technical language to interpret and to produce representations of the object, and the use of this knowledge in the construction of an object.

At levels 9–10 of secondary school the program is based on a practical project which includes reflection on the processes involved and the acquisition and comprehension of relevant concepts, abilities and attitudes. Through the completion of projects the program offers opportunities for the students to design, execute and evaluate activities that suitably represent the four phases in the life-cycle of a service: 1) analysis of requirements, 2) design, 3) production and development, and 4) communication and impact analysis.

The study program in technology education at grade/levels 11–12 is not compulsory, (as it is from levels 1–10) and is designed to provide students with a coherent and systematic vision of productive organizations and finally, integration into the workplace. The relationships between productive organizations and the social and natural environments, as well as the internal modes of operation of these organizations, are analysed. At these levels of study, basic management tools are also developed by the students (Elton, 2002).

As a result of the progress made in the field of technology education, and especially with the inclusion of technology literacy from K–12, there is now a much higher level of student awareness of the technological world. Students have become conscious of technology processes and products created by people for people, resulting in their gaining a deeper understanding of the urgent need for the development and application of their creative talents and capacities.

With respect to the second problem mentioned above, that is, obsolete and inefficient study programs to certify technicians, the programs to train technicians have mainly focussed on the traditional categories such as electricians, welders, mechanics, carpenters, sales personnel, cashiers and secretaries, of which there were over 400 before the Educational Reform, have now been reduced to only 46 categories. Restructuring of the study programs and the long overdue reduction in the large number of technician categories have resulted in significant progress in adjusting to the larger national and international trends in technological and industrial development. It has also made it possible for Chile to expand its vision beyond merely catering to the circumstantial and immediate demands of domestic employers. Now there is a new broader perspective regarding the long term requirements of Chile's industrial needs and potential.

In an attempt to remedy the third mentioned problem, which is obsolete equipment, tools and instruments, contents and methodologies, the Ministry of Education has developed and established an extensive project for the modernization of vocational secondary education and especially professional technical education, entitled *Chile Qualifies*, which combines the technical and consultative resources of several government ministries and services. Among other changes, new methodologies such as *Dual Education* (apprenticeship training), are being developed and applied. Dual Education is a combination of on-the-job training and classroom studies over an extended period of time. Another change in the methodologies has resulted in the development of the Modular Curricular Approach. This methodology divides the contents of a program into specific study areas in order to establish a coordinated and progressive system of study leading to a minimum professional qualification. Upon completion of this modular study program the student can either enter the work force or continue onto the next study level. Another improvement is the development of an extensive project to define performance standards for each specialty in technical education.

These important changes have resulted in several major modifications in the Chilean educational system, particularly in technology teacher education.

HISTORY

To understand the concept of education in contemporary Chilean society, it is necessary to view it against the historical model which prevailed before the military regime. Previous to 1973 the Chilean government tended towards a socialistic economy with a centralized political structure. Following the military take-over these were replaced by a strong central government and the development of a social market economy. The change to a market economy resulted in drastic changes in the educational system and its policies for development. The most outstanding of these changes, which took place in the 1980's, were: 1) the decentralization of the educational system which resulted in the transfer of the management of primary and secondary schools from the central government to the local municipalities or city councils, 2) the government changed the salary payment system from a direct-to-teacher arrangement to a grant-in-aid system based on student attendance, and gave the municipal governments the authority to manage this payment system, 3) primary and secondary teachers' status as public employees was eliminated, 4) in higher education the two traditional public universities (University of Chile and the Technological State University) were subdivided into 16 new universities, including two pedagogical universities, 5) the creation of a three-level tertiary educational system: universities, professional institutes and technical training centres, and 6) the initiation of the development of private educational institutions (Brunner, 1999).

In the 1990's, while still in the process of re-establishing democracy, the new government put special emphasis on curriculum reform with the objective of providing and assuring educational quality and equity for the entire population. However, the new government did not institute changes in general educational policies except to change the compulsory attendance law from ten to 12 years. These 12 years of education are now compulsory for all Chilean citizens aged six to 21 years. This 12 year mandatory attendance amendment was added to the National Constitution in May of 2003 and became effective in 2004.

In higher education, the Constitutional Law of Education, passed in 1990, established the Licentiate Degree (comparable to the 4 year Bachelor's Degree) and stipulated that it be granted only by universities. Obtaining the Licentiate Degree was a requirement before one could acquire official certification in the 12 traditional professions, for example medicine, architecture and engineering. Later, modifications to this law incorporated several more professions. Teacher education was incorporated in 1994, but did not include teacher education for the fine arts and technology education.

Primary school teachers in Chile had been educated in so-called Normal Schools from 1842 to 1973, when the military government closed the Normal Schools. Since that date, teacher education has been undertaken in public and private universities. In 1889 the Pedagogical Institute was created in the University of Chile. Since then, the University of Chile has been the only institution responsible for educating secondary school teachers. A few years later the new universities (including Universidad Católica de Chile, Universidad de Concepción, Universidad Austral de Chile, Universidad Técnica del Estado) initiated a secondary teacher education program (Núñez, 2002).

Until the creation of the Technical State University (TSU) in 1946, technology education was not considered an educational priority. The TSU initiated technology education and then technology teacher education in Chile in the Technical Pedagogical Institute which was created a few years later. Unfortunately, in 1976 the Technical Pedagogical Institute was transformed into the General Studies School and lost most of its pedagogical functions. Subsequently, in 1981, the TSU was transformed into the new University of Santiago (Muñoz, Norambuena, Ortega, Pérez, 1987).

However, most of the 16 regional universities which were born out of the University of Chile and the Technical State University, all of which are located outside of Santiago, continued to develop teacher education programs and one of these regional universities (University of Atacama) continued training technicians in their professional institutes. Then, in 1993, the Metropolitan Technological University was created in Santiago, and presently trains industrial engineers and technicians for several fields.

At present the Chilean government, impacted by the new requirements of the Free Trade Agreements, acknowledges the urgent and critical necessity for technician and technologist training in advanced technology and the need for teachers to teach in these areas.

OVERVIEW OF TECHNOLOGY TEACHER EDUCATION

Currently technology teacher education has three tracks: 1) teacher education for technology literacy taught in K–12, 2) teacher education for vocational education in secondary technical schools in levels 11–12, and 3) technology teacher training which prepares teachers to teach in technical training centres, professional institutes and/or university technology courses such as electronics and computer science. However, only completion of the first two tracks is obligatory in order to teach in K–12 and vocational education. As a result, teachers for professional institutes or universities now need only a higher technician certification. Nevertheless, most universities in the country currently offer pedagogical instruction programs for these teachers.

It should also be mentioned, with respect to technology teacher education, that the Chilean system of centralized education results in a high degree of homogeneity in the teacher education programs throughout the system. The institutions that offer programs in teacher education tend to use very similar pedagogical approaches in all the specialties, both in programs for teacher certification in technology literacy education (K–12) and for vocational teacher education (levels 11–12).

In the mid 1990's the Ministry of Education realized that the reform effort would be hampered if the quality of teacher education did not improve. However, the Ministry of Education did not have a clearly defined category in which to include teacher education nor the authority to enforce changes in programs administered by autonomous universities. Therefore, in order to generate interest in creating a more innovative environment in this educational area, the government decided to offer monetary incentives to reward innovative project proposals presented by institutions wishing to improve their four year teacher education programs. A sum of US$25 million, as a competitive stimulus, was allocated for this purpose. The winners of this award, 17 universities, implemented important modifications in the philosophical, methodological, administrative and material aspects of their teacher education programs in all specialties, including programs for technology teacher education, K–12 (Avalos, 2002). Fortunately, these modifications have been adopted by other institutions, even though they did not qualify or compete for the award. A synthesis of the most important innovations and modifications that have also been included in technology teacher literacy and vocational teacher education are as follows:

- A change to a more modern educational paradigm.
- An increase of opportunities for independent learning and the use of varied sources of information and a reduction in the lecture format as a teaching method.
- The creation of curricular activities which lead to the development of more critical and reflective intellectual attitudes.
- The intensive use of computers, multimedia and internet in learning and teaching, including communication and information technology.
- A curriculum reorientation towards more creative and progressive laboratory, classroom and field techniques (including apprenticeship training).
- A change in the approach to curriculum evaluation using standards, benchmarks, rubrics and portfolios.
- The addition of real problem resolution techniques or Problem-Based Learning (PBL) at some universities.

These important curricular modifications have been implemented in many of the major and minor courses of study at all levels of teacher education, and have brought Chile somewhat up-to-date in terms of its understanding of the need for constant change, development and growth in this field of education.

Even though the curriculum changes in technology education (K–12) were established before the initiation of the first programs in teacher technology education (at the beginning of 2000), the newly developed programs that were created in 2003 have already been implemented and many of the changes listed above are already in place. For example: the amount and quality of pre-service field experience; the standards for assessment of expertise and procedures among novice and advanced practitioners; intensive use of computers, special computerized equipment, and multimedia and Internet in learning and teaching. Also implemented are the curricular activities which lead to the development of more critical and reflective intellectual attitudes including the Problem-Based Learning model, and outside participation from the industrial technology sector in curricular activities for technology teacher education.

STRUCTURE OF TECHNOLOGY TEACHER EDUCATION

The structure of teacher education in Chile is as follows:

1. Primary school teacher certification to teach grades 1–4 requires eight semesters of study. However, to teach grades 5–8, students can elect to study a minor by concurrently taking a specialization such as mathematics, language or science.

2. Secondary school teacher certification requires 10 semesters with a major in mathematics, the sciences, arts, or a foreign or native language, which must be taken concurrently. In some universities, the Licentiates who have graduated, generally with 10 semesters of their major area, become qualified pedagogically after two or three semesters, obtaining the necessary qualifications to teach from grades 9–12. These consecutive secondary teacher education programs are currently offered at four universities, but they have small numbers.

Technology Teacher Education in Chile

Currently there are two categories of technology teacher education programs, one for technology literacy and the other for vocational secondary education. The first category includes a cycle of three different study programs that lead to certification in technology literacy education, for teaching in the primary and secondary humanistic-scientific schools: first cycle (1–4), second cycle (5–8) and third cycle (9–12). The second category includes a study program that leads to certification for teaching professional technical education in secondary vocational schools (11–12), and occasionally for teaching in the technical training centres and/or professional institutes.

The certification required for teaching technology literacy in kindergarten and the first cycle in primary school (1–4) can be obtained while studying to be a pre- or primary school teacher. The certification for teaching in the second cycle of primary school (5–8) requires the completion of a course of study with a minor program in technology education which must be taken concurrently with the major in primary teacher education. The certification for teaching at the secondary vocational schools (9–12) requires three or four semesters of study after having obtained the technician certification at a tertiary institution.

Technology Teacher Education for Primary and Secondary Education

As a result of the new educational reform, technology education is now included in the primary and secondary curricula and a few universities now offer training programs for teachers in this field of study. At present, there are four study tracks leading to certification in technology teacher education. The requirements for graduation are as follows:

1. Completion of three or four semester courses within the curriculum for primary teacher education and training in teaching methodologies to teach technology education to levels 1–4. The study plan to obtain this certification includes four semester courses of technology education in special education teaching methodologies and its practices; for example: Foundations of Technology Education during semesters III or IV, Technology Education Teaching Methods in semester V, and Technology Education Professional Practice in semester VIII (Figure 1).

Figure 1.

Primary Teacher Education (Technology Literacy Education, Concurrent)	I	II	III	IV	V	VI	VII	VIII	IX
			(1)	(2)	(3)			(4)	

2. Completion of four or five semesters of study concurrent with the pre-graduate program at the university. Concurrent study in a minor is necessary while completing the course of study leading to the primary teacher education certification to teach technology education at levels 5–8 (Figure 2).

Figure 2.

	I	II	III	IV	V	VI	VII	VIII	IX
Primary Teacher Education (Major)	I	II	III	IV	V	VI	VII	VIII	IX
Specialization (Minor) in Technology Education					V	VI	VII	VIII	IX

Students select only one specialization (minor) to be taken concurrently at the beginning of the fifth semester in primary teacher education and Licentiate (Bachelor) in Education. Some universities require an additional ninth semester when there is a minor.

3. Completion of the minor requires two or three semesters of study after post-primary teacher certification at a university or professional institute. This certification is required in order to teach in technology education at levels 5–8 as in track 2. Most training programs for primary technology teacher education include three areas of study: fundamentals of technological education; methodologies, techniques and handling of instruments for technology education; and teaching practice, which takes place at primary schools with the supervision of a university faculty member.

4. Finally, completion of the major requires a specialized program in technology teacher education of ten semesters at the post secondary level which leads to certification for teaching in technology education in levels 1–12. In a few universities the completion of these same study requirements can lead to a specialized teacher training certification for teaching specific courses such as electronics, mechanics, and wood & metal working. This certification qualifies the teacher to teach these courses at the post-secondary level in professional-technical education or in the secondary vocational school study programs at levels 11–12. This certification is currently offered only in two universities in Chile.

Current Problems in Technology Teacher Education (1–12)

Even though there are a variety of study programs leading to technology teacher certification in primary and secondary school, student enrolment is critically insufficient to meet the demand. This problem has been recognized by the Ministry of Education, but has not yet been resolved.

At the moment, the country does not have enough qualified teachers in the field of technology education. Two different studies, one composed of 145 primary teachers and another with 98 secondary teachers across the country, reveal that the technology teachers come from a variety of disciplines, which are in decreasing order: crafts, technical professional specialities, natural science, social science, mathematics, languages and physical education. These teachers have become involved in this field primarily because school administrators have given them no other choice. A very small number have a genuine interest in teaching in the field of technological education.

The Ministry of Education is currently offering an in-service teacher training program for teachers who are willing to accept the challenge of working in this area. However, in reality little can be done about the shortage since the universities in Chile do not offer enough programs in technological teacher education to meet the growing demands. The Ministry of Education has made two public calls in hopes of finding candidates: in the first call only 20% of those who applied for enrolment into the study program were admitted, in the second only 30% (Elton, 2002). Table 1 demonstrates the present situation.

Table 1. Number of Programs in Technology Teacher Education (Literacy)
(Source: Information provided by the Institutions.)

Technology Teacher Education Tracks	University	Professional Institutes	Total
a) Ten semesters pre-grad study	2	0	2
b) Two/three semesters of study along with in-service training courses at universities and/or professional institutes	2	3	5
c) Four/five semesters concurrent in general teacher education study at universities	2	0	2
d) Two or three semesters of study in the fundamentals of technology and teaching methodology courses concurrent with study in primary teacher education	14	0	14
TOTALS	20	3	23

In 2003 there were 61 institutions of higher education with programs in general teacher education. Of these, 53 were universities and 8 were professional institutes (Ministry of Education. 2003). Of these, only 23 institutions offered programs in technology teacher education in one or more of the various levels in primary and/or secondary education.

In 2004, approximately 11,000 students enrolled in primary teacher education programs. Of these, only 261 students are currently studying to teach technology education at levels 5–8 (see Table No 2). This disparity takes on serious dimensions when we consider that there are currently 8000 primary schools which urgently need teachers trained in technology education. The statistics in Table 2 further demonstrate that, at the moment, technology teacher education for primary and secondary education in Chile has not effectively adapted to the changes produced by the Educational Reform. Only two universities train specialists in technology education within their regular study programs. To make matters worse, at present less than half of the institutions train their students in the foundations of technology education and the necessary methodologies required to teach in the first cycle (1–4) of primary school. In the next five years more than 2000 specialists will be needed to teach in the area of technology education in Chilean primary schools. This need will force the Ministry of Education to offer urgent teacher training programs and propose an increase in programs for technology teacher education in the universities and professional institutes.

Table 2. Number of students enrolled in Technology Teacher Education programs for Primary Education in Chile 2004.

Types of Programs	Universities	Professional Institutes	Totals
Programs for first cycle (1–4)	4206	470	4676
Programs for second cycle (5–8)	143	118	261

Technology Teacher Education for Secondary Vocational Schools

The traditional division between general and vocational secondary education, where students started their vocational specialization at the age of 14, was also modified by the curriculum reform in 1998. This change extended the general education curriculum to 12 years (primary and secondary education) for all students. The last two years of secondary education (11–12) are now organized into two different modalities: scientific humanistic education (56%) and vocational education (44%) representing an increase of about 10% in vocational education (Cox & Gysling, 1990).

Teachers for secondary vocational education (11–12) become pedagogically qualified after two or three semesters of tertiary study. To begin these studies they must first complete and acquire their technician certification for which they study at post-secondary institutes or universities. This same certification that is used for teaching levels 11–12 also allows the teacher to teach at the technology centres.

At present, education for teaching at vocational schools and occasionally in technical centres and professional institutes is a difficult problem, particularly because most universities are not sympathetic to the specific training needs for those students trying to qualify in the field of technology teacher education. As a result, only five state universities and three private universities (of the total of 65) are now offering technology teacher education programs for secondary vocational schools. Fortunately, the Chilean government has instituted a new project in the area of technology education and the development of technicians entitled *Chile Califica*. "The team (OECD) understands that as part of the *Chile Califica* programme tenders will be called soon for the development of an appropriate training programme for vocational teachers. This is a positive step but it will need to be followed up by the selection of those universities that are to offer this programme" (OECD, 2004, p199).

AN EXAMPLE: UNIVERSITY OF ATACAMA

The School of Humanities and Education at the University of Atacama now offers study programs in technology teacher education at the primary (levels 1–8) and secondary vocational schools (levels 11–12).Within these study programs for the primary teaching certificate the student also obtains certification for teaching levels 1–4 in technology education. For levels 5–8, a Technology Teaching Certificate is obtained through a specialized study program, (a minor area of study) which is now part of the required coursework leading to the four year degree in education (comparable to the Bachelor's Degree) and the Primary Teaching Certificate. The Secondary Vocational Teaching Certificate for teaching technology education is obtained through four semesters of pedagogical study after obtaining the technicians diploma.

Problem-Based Learning (PBL) in Technology Teacher Education

Since the year 2000, the School of Humanities and Education at the University of Atacama has been implementing a new curriculum in teacher education based on a model known as PBL. This curricular approach has been used in the training of physicians, nurses and engineers in the United States and other countries, and it is now being applied in teacher education in Latin America (Iglesias, J. 2002). The PBL educational concept is considered by many theorists to be the most effective means for applying the principles of constructivism in the learning process. Savery & Duffy (1995) have pointed out that PBL, as described by Barrows (1985, 1986 and 1992), possibly represents the best example of a constructivist learning environment.

Wilkersonn & Gijselaers (1996) see PBL in the context of an approach to learning rather than as a teaching technique and feel that the traditional curriculum overloads students and complicates the learning process by placing an excessive emphasis on memorization. Moreover, they see PBL as a means to empower students in the development of their problem-solving skills, rather than a means for learning simply for the sake of acquiring knowledge. This vision is in line with Barrow's analysis (1986) which associates PBL with a particular learning strategy designed for working in small groups with a tutor. This method, Barrow pointed out, is consistent with the principles of learning in adults and makes it clear that courses must not only educate well, but should also provide the fundamental basis for a process of formal and informal life-long learning.

The curriculum for teacher education at the University of Atacama uses a PBL approach in two forms. One form is a PBL module required during each semester for the students with a major in technology education. Regularly scheduled classes are cancelled for all pedagogy students and during this time the students work in teams of approximately ten members with a tutor in trying to solve real problems, such as the student dropout rate, use and care of personal computers, violence in schools, drug and alcohol addiction, and specific student learning problems. In the second form, the minor in technology teacher education, which is organized in modules, includes problems that are solved by use of computer assisted software and by applying the PBL techniques. The PBL problems designed for these subjects in the minor are based on real life problems which are encountered in all of the subject areas and especially in each technology module, and are used as the initial motivators. Problems determined for each module make it necessary for the student to design, assess and construct one specific technological solution to the given problem. This technology teacher minor includes the following modules: computer graphics and animation, information technology, laser technology, pneumatics, electricity and electronics, material processing, constructional technology, structural engineering, transportation technology, energy and power research and development.

Due to the spontaneous and unstructured nature of these problems (at least ostensibly) there is, of course, no specific pre-programmed learning or logical thematic study sequence that can be applied in the learning process. This is because they are real life problems and are therefore unpredictable. In some cases, however, problems arise spontaneously during the course study in the minor or the major which provide a very good opportunity to apply the PBL techniques for problem solution.

The minor program consists of a total 704 hours of study within a two-year period and is integrated into a major curricular system. The coursework in this specialized study program is completed during the last two years of study leading to the primary teaching certificate. The core of this specialized program is composed of activities that are completed in what is called the *Technology Education Collaborative Laboratory* which provides an interactive and collaborative environment where the PBL methodology can be discussed and applied. This collaborative laboratory is a classroom especially equipped with interactive software, internet, learning materials, tools, instruments and equipment and specially equipped work stations. All are focused on various technological themes (12 modules) which can be studied and experienced hands-on within a collaborative and interactive framework. Teams of two or three students rotate through the various modules and work stations, participating in activities specifically designed to help them develop their skills and strategies, and draw out their potential to creatively resolve technological problems.

During the second year of the specialized study plan, students must complete their professional practice and their seminar project, including research, in some area of technology education. Once they have completed their professional practice, research seminar and coursework, they may graduate and will receive their primary teaching certificate with a minor in technology teacher education. This enables them to teach from levels 5–8.

TEACHER CERTIFICATION

In order to teach in any school, and in any speciality, it is mandatory in Chile for all students to obtain a teaching certificate from an officially accredited and authorized university or other institution of tertiary education known as a professional institute. The universities as well as a few professional institutes are legally authorized to grant certificates to teach in primary schools, for the first cycle (1–4) or the second cycle (5–8) or for secondary schools in a specific knowledge area.

Nevertheless, in agreement with the Constitutional Teaching Law (LOCE) the authorized professional institutes can provide technology teacher education programs and grant the appropriate certification required to teach technology education in secondary vocational schools, but not in kindergarten or primary schools.

It should also be born in mind that the Ministry of Education can grant temporary authorization to other professionals or high school graduates to teach at any level in the educational system when there are no properly qualified applicants interested in the positions available. However, they must obtain the required teacher certification within a specified period of time.

The certification to teach technology education in primary schools is obtained after an eight semester program of university study with a major in primary teacher education. This certification can also be obtained by completing three or four semesters of study at university in technology teacher education after obtaining the certification to teach in primary schools.

The certification to teach technology education in secondary vocational schools is obtained after a three or four semester study program at a university or a professional institute, and after obtaining the technician certificate from a tertiary educational institution. Practically all of these programs are offered as evening classes to facilitate those teachers who are working in schools without certification.

CONCLUSION

The new Chilean Educational Reform now includes a literacy in technology education program with an updated and modernized focus.

However, in practice, there have been very few, if any, productive results. There is still a very serious shortage of properly trained technology teachers, and the majority of the universities and professional institutes, that have education departments and teacher training programs, do not have technological literacy teacher education programs. As a result, technology courses in the majority of schools are taught by teachers who have little or no interest and no training in teaching technology education; for example, language, art, history and physical education teachers are usually asked to teach these courses, and these teachers very reluctantly accept. Therefore, of course, most Chilean secondary students do not get very excited or motivated about studying for a career as a technician (such a mechanics, mining equipment maintenance or domestic electronics).

This type of technology education, centered on the development of manual expertise (often using obsolete technology), has created in children and youth the perception that studying to become a technician does not require any significant level of intelligence, and that being a technician is only for those students who lack the necessary intellectual skills to pursue university studies. It is also thought that these careers are for students who come from families at the lower income levels, who cannot afford university tuition and expenses, or who need their recently high school graduated son or daughter to start working immediately to help with family expenses. As a result, those studying to become technicians are generally stigmatized with the negative image of being socially deprived and/or intellectually inferior.

The dilemma that Chile currently faces with respect to this issue is that the government has recently signed free trade agreements, buy/sell arrangements and market contracts with several countries in various parts of the world. It is also evident that this process of establishing market, finance and trade relationships with other countries will continue to increase in the future. These countries, with which Chile is negotiating, will demand a high level of technological development, skill and expertise. The country's success in developing, maintaining and expanding its international economic and trade relationships, and its ability to inspire confidence in its capacities to produce and distribute quality goods and services, will be directly determined by the extent to which Chile is prepared to meet the above demands. This, of course, makes it urgent that Chile creates and implements more programs with well-qualified teachers in technology teacher education.

As previously mentioned, in 2002 the government established an extensive project for the modernization of professional technical education; this project is entitled *Chile Qualifies (Chile Califica)*. While the program has been very effective, if it does not begin attracting better students it will not achieve the hoped for results. This means that the Ministry of Education in Chile must begin immediately to develop and promote innovative and imaginative training programs in the field of technology literacy education. If Chile fails on this account, students will continue to prefer the traditional professions rather than the technical professions. Chile needs to begin creating in its best students the desire to study technology as it has in the past in the traditional fields of study.

Experts in the Chilean Ministry of Education are very aware of and concerned about these problems and difficulties, and are initiating an intensive campaign to resolve them. In this campaign lies the hope for more effective and successful Chilean technological growth and development.

REFERENCES

Avalos, B. & Nordnflycht, M. E. (1999). *La formación de profesores. Perspectivas & Experiencias.* Santiago, Chile: Editorial Santillana.

Avalos, B. (2002). *Profesores para Chile:Historia de un proyecto.* Santiago, Chile: Ministerio de Educación.

Barrow, H. S. (1985). *How to design a Problem-Based Learning Curricula for Pre-Clinical Years.* New York: Springer Publishing Co.

Barrow, H. S. (1986). A Taxonomy of Problem Based Learning Methods. *Medical Education,* 20, 481–486.

Barrow, H. S. (1993). *A Problem-Based Learning in Secondary Education and the Problem-Based Learning Institute.* Monograph. Springfield: Southern Illinois University School of Medicine.

Barrow, H.S. & Kelson (1996). A Problem-Based Learning and problem solving. *Probe Newsletter of the Australian Problem-Based Learning network.* 26, 8–9.

Barrow, H.S. (1996). Problem-Based Learning in Medicine and Beyond: A brief Overview. *Bringing Problem-Based Learning to Higher Education: Theory and Practice* . 68, 3–11.

Brunner, J. J. (1999). *Informe e Indice sobre la Capacidad Tecnológica.* Santiago, Chile: Universidad Adolfo Ibáñez.

Cox, C. & Gysling, J. (1990). *La Formación de Docentes en Chile 1842–1987.* Santiago, Chile: C.I.D.E.

Elton, F. (2002). *Technology education in the Chilean curriculum reform.* Presentation in Technology Education Seminar Holland. (National Coordinator of Technology Education). Chile: Ministry of Education.

Iglesias, J. & Vera, R. (2001). Problem-Based Learning in Initial Teacher Education. Program at the University of Atacama. *International. Year Book of Teacher education, 97–98.* Word Assembly. Santiago, Chile.

Iglesias, J. (2002). Problem-Based Learning in Initial Teacher Education. *Prospects,* XXII(3), 319–332. Paris: UNESCO.

REFERENCES

Iglesias, J. & Sills, N. (2004). *Initial teacher training in Technology Teacher Education at the University of Atacama.* Unpublished dissertation. Project ALFA RIUFICEET. Marsella, Francia: UniMéca.

International Technology Education Association. (ITEA) (2000). *Standards for Technology Literacy. Content for the Study of Technology.* Reston,Virginia: Author

Ministry of Education Chile. (2000). *Chilean Curriculum Reform (1996–2002).* Santiago,Chile: Unidad de Currículo y Evaluación.

Ministry of Education Chile. (2002). *National Chilean Inform for Centre for Co-Operation with Non-Members. European Community.* Santiago. Chile: Unidad de Currículo y Evaluación.

Muñoz, J., Norambuena, C., Ortega, L., & Pérez, R. (1987). *La Universidad de Santiago de Chile.* Santiago.Chile: Universidad de Santiago de Chile.

Núñez, I. (2002). La formación de Docentes. Notas Históricas in Avalos B. (Ed), *Profesores para Chile.Historia de un Proyecto.* Santiago.Chile: Ministerio de Educación.

OECD. (2004). *Reviews of National Policies for Education. CHILE.* Paris. France: OECD Centre for co-operation with non-members.

Savery, J.R. & Duffy,T. (1995) Problem-Based Learning: An Instructional Model and its Framework. *Educational Technology,* 35(5) , 31–38.

Soto R, F. (2000). *Historia de la Educación Chilena.* Santiago, Chile: C.P.E.I.P.

Wilkersonn, L. & Gijselaers, W. (1996). Concluding comments. In *Bringing Problem-Based Learning to Higher Education: Theory and practice. New directions for teaching and learning,* 68, 101–104. San Francisco USA: Winter Jossey-Bass Publishers.

Technology Teacher Education in France

Chapter 3

Jacques Ginestié
Professeur Des Universités, France

INTRODUCTION

General Background of French School System

In France, school is compulsory for all children between the ages of five and sixteen. However, infant schools[1] have an obligation to provide nursery care for children from the age of two, and no students can leave school before the age of eighteen unless they enter vocational training or find a job. The legal obligation in fact becomes an obligation for people to be in education from age three to eighteen. The French school system is organised at two levels; primary education and secondary education.

Primary education involves two schools, the infant school and the elementary school, which in turn are divided into three stages: the initial learning stage for children aged two to three and up to five years; the basic learning stage for children from five to eight years and the fundamental stage for eight to eleven year-olds. Secondary education starts with four years schooling at the lower secondary school (College) for all the 11–16 year-old pupils. Upper secondary school (Lycée) offers three options to the 16–18 year old pupils: the general option, with literary and scientific subjects; the technology option with industrial, service and biotechnology subjects and the vocational option, which covers all the vocational sectors. The Baccalaureat examination concludes these secondary studies and in the year 2000, approximately 70% of French pupils reached this level of education.

This broad access to secondary education is accompanied by a redefinition of the aims and objectives of each of the education levels. Generally speaking, schooling up to the end of the secondary stage fits into the context of general education for all, whereas the objective of the Lycée is to guide pupils towards a fast track vocational education via the various streams of the vocational Lycées or towards university education via the general Lycées.

---

[1]With a national curriculum definition

One of the major debates about education for all concerns equality, and is a recurrent theme, inscribed in the republican values of secularity of schools and free education[2]. General education deals with an equal access to culture, citizenship, and social integration while vocational training deals with the equal opportunity to access a qualification and a profession. As with general education, the French Ministry of Education organises and controls vocational education. Each level of qualification corresponds to a diploma delivered by this Ministry. In fact, there is a strong link between general education and vocational training: the pupil's orientation is a central question in the French education system. The pupils' choice is determined by their results in general education and by the opportunities in vocational areas; job qualification descriptions and the number of opportunities provided by the labour market. Some institutes (CEREQ[3], CNP or DEP) independently describe the professional qualification, assess the needs of the labour market and develop the curriculum. The main problem is the disconnection between the possibilities of training offered and the young people's wishes in terms of professional orientation. Evidently, these elements have consequences for school organization, for the repartition between the subjects, and for the pupils' assessment and the curriculum, particularly for Technology Education (TE).

Organization of School and Curriculum

Primary School

At primary school, technology education is combined with science education in three stages: *discover the world* at the first stage (3–5 years old), *discovery of the world* at the basic stage (5–8 years old), and *scientific and technological initiation* at the fundamental stage (8–11 years old). This precedes subjects such as *technology, physical sciences* or *biosciences and geosciences*, when students arrive at the college aged about 11 years. In this system, children pass between these areas through the educational activities when they experience the ability to observe, to manipulate, to experiment,

---

[2]In France, 80% of the children attend state schools and 19%, private schools under contract (mainly Catholic), i.e. schools that undertake to offer the same teaching programmes, observe the same school organisation (hours, objectives, examinations and evaluation, etc.) in return for which the government takes responsibility for their budget and personnel costs, which are calculated on the same basis as those in the public sector.
[3]CEREQ : Centre d'Étude et de Recherche sur l'Emploi et les Qualifications ; CNP : Conseil National des Programmes ; DEP : Direction des Études et de la Prospective

to make, to manufacture, to design... This organising principle progressively structures knowledge by giving meaning to the subjects. Therefore, the discipline matrices can be defined via the principles of a specific curricular organization constituted to form a whole (Develay, 1992). This form of organization has to be assessed in relation to teaching and learning issues (Charlot, 1997). The process of subjects' delimitation encourages a constructivist approach, whereby the pupil learns new knowledge as the result of the discovery activities conducted in class. The pupil does not organise it in terms of pre-defined school subjects, but according to unifying criteria, such as the nature of the knowledge handled, the action it allows or the methodologies used in the activities (Ginestié, 1999). From this viewpoint, school is a transition from a sensitive and intuitive perception of the world to a rational relationship via knowledge constituted as school subjects.

At primary school, teachers use one classroom in which they have to teach all the subjects. They focus mainly on the teaching of French and mathematics, called fundamental learning. Some teachers attempt to teach science and technology education. Among those, the majority teach natural sciences through observation of life development (plants or small animals like red fish or mice); a few develop some electricity experiences and fewer still try to implement some technology education. Technology education is widely interpreted as handicraft activities or applied science (building a torch after the study of electrical circuits), and is called technology because of the use of some skills and materials such as scissors, cutters, cardboard or glue. The Ministry of Education tried to develop science and technology education through two initiatives (PREST and Lend a Hand); but the results were not really satisfying, mainly because of inappropriate teacher training.

The Lower Secondary School: The College

At secondary college, technology education appears as a subject for the four years. The subject was introduced in the early sixties but the generalisation for all the students at the college in 1985 represented a very important change of curriculum orientation. This initial curriculum changed again in 1996. It is organised on the basis of competencies that pupils must acquire and activities to achieve these competencies. The four years at the college is a progressive construction, with the last year being a synthesis of the three previous years. Technology education at the college is based on the industrial project method. This method is a process by which an enterprise develops a new product for the market and proceeds through all the stages from the first description of the new product to its distribution. During one and a half hours per week for the three first years, pupils experiment with all the different scenarios of the industrial method corresponding to the different stages (needs analysis, design, industrialisation, production, distribution, commercialisation, use…). For two hours per week in the final year, pupils organise all these different elements in conducting an industrial project by themselves. Through this realisation the pupils can develop a mastery of technical choices, material solutions as well as working through methods (Hill, 1998). In this way, TE is taught through two perspectives:

- The macroscopic perspective presents industry as the place of design, manufacturing, control and commercialisation of technical objects. In this perspective, TE approaches industry as a very complex technical system of production. Understanding this system requires identification of the different organization principles (structural, functional, hierarchical…), the orientation techniques used and the various jobs. Analysing this system can be done by decomposing the sub-systems and studying the interactions between them. The complexity of the possible architectures (number of significant sub-systems and social practices possible for each of them) is rarely studied at school. Nevertheless, the architectures are significant social organizations and examples of practices of industry.

- The microscopic viewpoint presents industry through the study and realisation of a particular technical object. This realisation simulates what happens in industry by the nature of activities, organizations, skills and tasks. In reality, pupils are constrained as the technical object they have to develop is defined by the teacher, the different technical solutions are pre-defined, and materials are limited. In fact, the pupils' logic becomes more an organization of successive school tasks without the real possibility of significant choice (Ginestié, 2002).

Knowledge about these industry organizations provides vocational knowledge for students, and so helps them make appropriate vocational educational choices.

The Upper Secondary Schools: The Three Lycées

TE appears in a different form according to the goals of each kind of lyceum (lycée): as an option at the general lyceum (3 hours weekly), as a main compulsory subject at the technological lyceum (3–6 hours weekly) and as an area connected with professional skills and knowledge at the vocational lyceum (6–9 hours weekly). The nature, goals and organization is different at each kind of lyceum. For the first, TE is a general subject designed to develop the pupils' relationship with the technological world. For a large part, these pupils will never choose a technological study or career; they study TE to develop their own general literacy in this field. For the second, TE is also a general subject but as an introduction to university vocational training, that is, to become an engineer or technician. For the third, TE is directly linked with the professional area; for example mechanical technology, components technology, biotechnology, or any other specialised area or technology. The goal is to give young people the necessary technological background to understand the professional skills and context.

Vocational training in France for the lower qualifications (as workers, office employees…; corresponding to levels VI and V in Figure 1) is based on three types of education: general study to give a broad background (general culture and professional culture), vocational training in companies by alternating training and extended periods of study, and vocational training at the lyceum. The aim is not to train specialists for a specific job but to provide a wide background as a base for professional and occupational flexibility.

HISTORY

After the Second World War, the education system in France was structured to organise the relationship between general education, professional-oriented education, the level of qualification and, at the end, employability. This organization was to be representative of the companies' structure with a wide range of qualifications and professional domains. Most of the different jobs and almost all the qualification levels are prepared through vocational training as part of the initial profession-oriented education. This structure of the school system was representative of the work social division. As shown in Figure 1, the entire school system was organised to provide paths to all levels of qualification. Teachers were selected on the double criteria of professional domain and qualification level. During this period, TE did not really exist as a subject in general education. There was a subject called Handicraft Education, with activities that were largely gender based. Some teachers had experimented with different forms of TE, but it was more like applied science or an industrial approach with a preference for mechanical construction (draft, design and manufacture).

Figure 1: General organization between general education, profession-oriented education and qualifications

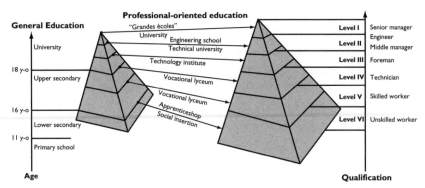

The economic crisis of the seventies and the eighties affected firstly the lower qualified jobs (level 5 and 6), and then progressively all the other levels with changes in the nature of the job, the structure of companies and the industrial organization. The education system was restructured with a distinction between general education and profession-based education (with general culture as the main reference for adapting to social and professional evolutions), the development of access to study for all, and extending compulsory schooling to 18 years.

Technology Education appeared as a general subject for all students at lower secondary school, firstly in 1976 as manual and technical education then as technology for all in 1985, and it also combined with sciences at primary school and as options at the lyceum. Because of high unemployment particularly in unqualified young people and the resultant social problems, the French government adopted, in 1989, the Education Orientation Law. With this law, France combined general and profession-oriented education, based on the goal of no young person leaving the school system without a qualification, apprenticeship contract or employment. However, the distance between this principle and the reality was great and its application difficult. But this law opened the possibility to develop TE both as general and professional education.

The same law changed the system of teacher training with the creation of the IUFM (university teacher training institute). For the TE teachers, training moved from training centres to universities, a change which promoted the interaction between the different components of university, specifically between training, development and investigation. As a result of these changes, the TE teachers profile evolved drastically at the same time: from being a specialist in a very narrow expert domain to the more general technology-educated teacher. Before this evolution, professional experience in enterprise was an important criterion (five years of experience in a company was equivalent to a Diploma of Engineering); at the end of the eighties and throughout the nineties, the most important criterion was the general education level in technology. The competitive entry examination to employ teachers favoured the university diploma rather than the experience.

OVERVIEW OF TECHNOLOGY TEACHER EDUCATION

The General Schema to Become a TE Teacher in France

To understand teacher training in France, two points are important. Firstly, a teacher in public employment needs to succeed in the competitive entry examination during the fourth year of university studies. Secondly, the course to become a teacher is based on a successive principle: students acquire competencies in the subject at university during the first three year cycle (licence) followed by two years at the IUFM during which they acquire the professional competencies (see figure 2). The first year at IUFM is entirely devoted to preparing for this competitive entry examination to French public employment. Each year, the French Ministry of Education indicates the number of positions available for each subject area throughout the country; for example, for the 2004 session, the Ministry proposed 185 positions in General TE and 90 positions in Mechanic Engineering. The level of competition between students for each subject area depends on the number of students; for example, for the same session, there were about 900 candidates for the 185 positions in General TE and about 300 candidates for the 90 positions in Mechanic Engineering.

Technology Teacher Education in France

Figure 2: Safety Organization

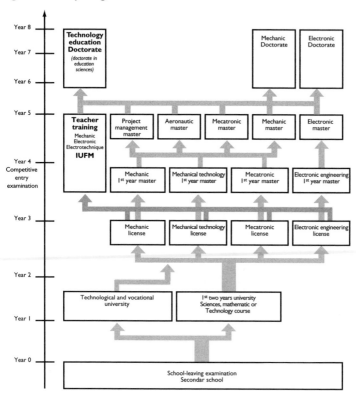

The First Degree in the Subject

As indicated in Figure 2, the initial university studies are organized to prepare for masters degrees rather than teacher training at IUFM. This represents problems for the TE teachers' recruitment. There is, in France, a general student disaffection for scientific and technological studies, which is a real problem when recruiting for teachers. This is a common concern with many European countries: young people do not generally elect this kind of study. For example, at Uniméca, Aix-Marseille University, in the technology area, the number of candidates who wanted to proceed with teacher training after their bachelor degree decreased by 40% between 1996 and 2001 (from 300 students to 180) while, on the other hand, the different masters' opportunities attracted 350 students. The second cause of the problem is directly linked to this situation but also with the general employment situation. At this level of study, students can choose between becoming an engineer, with a good level of income and a low rate of employment security, or becoming a teacher, with a high rate of employment security and a low level of income. If the socioeconomic situation is good, it is very difficult to recruit for teacher training. When the situation depresses (for example, an increase in the unemployment rate), there is an increase in the number of candidates who want to become teachers. The third general cause is due to the initial level of recruitment of TE teachers. In France, many TE students follow a short technological university cycle, of two years. They do not have an adequate level to proceed to the competitive entry examination and, for those who want to become teachers, the level of the last year of the bachelor degree is much too high and the requirements are too hard for them. Generally, they do not succeed and so they return to industry.

To try to solve this situation, different strategies have been developed at different levels. The first way is through general TE, but this is a vicious circle: numerous well-trained teachers are needed in order to promote TE; but, it is very difficult to find good teachers and there is little time for training them. In this context, many relationships and projects have been developed with primary and secondary schools through exhibitions, project activities, resource support and in-service teacher training. But this approach is a long term perspective and it is difficult to predict the effect. The second strategy introduces modules of technology teacher training during the first cycle at university. These modules have existed for a long time at university in the subjects of mathematics, literature, foreign languages or humanities, but not in TE. The third strategy has been to develop and to promote the accreditation of prior experience and knowledge. Since the orientation law, universities can exempt some students from prerequisites to entry at one level, according to their experience and its curriculum. This process is recent and has not really developed specifically for TE. Rather than a single innovative project, many strategies have been developed in different ways, to adapt to the need.

STRUCTURE OF TECHNOLOGY TEACHER EDUCATION

Lower and Upper Secondary School Teacher Education

The First Year of TE Teacher Training

The first year is fully devoted to preparation for the competitive entry examination, success in which requires a high level of performance. The nature of the examination prioritizes academic and theoretical knowledge and places less emphasis on improving knowledge, competencies or abilities to teach. In other words, to be a good teacher, in France, you need to master the subject you have to teach, but not really the ability to teach. This problem applied to all teacher training in France, regardless of the subject.

According to this orientation, there is a problem when the subject taught at secondary school does not correspond with the subject taught at university, which is the case for general Technology Education. There is no specific degree course that corresponds to the aims and the content of Technology Education. Students come mainly from mechanical and electrical degrees, but also from management and other areas. It is difficult to base the first year of the course on the knowledge of previous studies. As can be seen in Figure 2, the specialization is limited in Technology Education teacher training. One of the goals of this first year is to give the students some knowledge to enable them to understand Technology Education as a general culture and can be summarized in five general approaches:

- The technological installations user: the growing sophistication of familiar machines induced a new person-machine rapport. Technology Education has to give meaning to this use of technical objects. Students have to develop competencies for analysing the action/process of machines through functional analysis skills;
- The products buyer: the diversification of makes, models and ranges of products renders the buying of a technical object increasingly perilous. Very strong cultural and social pressure is strengthened by advertising messages based on wants and the temptation to ignore the price. Technology Education has to increase the understanding of technical objects' production and existence modes. The students have to develop knowledge about the interrelations between functions and values and also about the artefact semiology;

- The technical systems user: social organizations link around increasingly sprawling user networks of systems that strongly induce peoples' activities and the environment's evolution. TE must highlight this user's interaction with the social systems, through the use of technical systems. The system analysis approach constitutes a reference field for students.
- The social actor in the production system: humans occupy an active place in the social organization as consumers but equally by work and social contribution. The consumption cycle: work = remuneration = buying power = technical objects is largely the basis of modern western societies. This model is culturally and historically significant. Technology Education could significantly contribute to the understanding of this evolution by clarifying relationships between human and technical objects. The knowledge students deal with about labour organization (and labour's social division) in connection with some concepts of general economy, including but not limited to the company's economy;
- The citizen of the City: technical and social evolution is always presented as experts work for the common wellbeing. Examples to show that this confidence in experts is often misplaced are numerous. It is an aim for Technology Education to give background and understanding to this kind of debate. Students have to elaborate with analytical skills to examine a new situation according to the context.

Studies combine theoretical and practical knowledge in various fields. Some of these elements are introduced in the first year but there is a great diversity of student backgrounds. With a 550 hours' curriculum, students have three groups of lectures:

- Technical group: is based on the technical and scientific knowledge required to know and to manipulate in these basic domains with some adjustment according to the personal background of the students (about 200 hours). This group includes mechanical and electronic design and fabrication, general and enterprise economy, automated systems, robotics and computer sciences. All these elements are widely computer-based.
- Project group: supports all the tools, analysis, language and knowledge required to conduct an industrial project (200 hours). During this course, students have to make their own project that they present as part of the selective entry examination.

- Technology Education group: includes a study of the Technology Education organization in the French school system (150 hours): history of Technology Education, school organization, kind of activities, teaching learning process, materials, tools, resources and possibilities available to teach Technology Education. In this group, students undertake a period in a school and improve their mastery to use the equipment available in the classroom.

For this training year, the contents and nature of courses are strongly linked to the organization of the selective entry examination but, at the same time, each IUFM is under national assessment and the results provide a good indication about the students' performance. Including all the different options for Technology Education teachers for lower and upper secondary schools, the Ministry of Education in 2003 offered 1,010 positions when 5,349 people applied for these jobs (18.89%). For all the options of vocational training and for vocational lyceum, there were 13,455 people who applied for the 2,878 positions offered by the Ministry of Education (21.39%). In France, there are 30 IUFM, one for each region. But not all the IUFM's offer all options and all subjects; that is, there are 22 IUFM which prepare the general Technology Education for lower secondary school. To illustrate this, the IUFM of Aix-Marseille has registered 40 students to prepare for this option, and 28 students succeeded in both groups of tests. In theory, anybody who presents the required condition can sit the selective entry examination; it is not an obligation to be registered in an IUFM. But in fact, the great majority of the students who succeed came from an IUFM. This fact is more evident for Technology Education, probably because the knowledge about materials, tools, machines and models and the place of practical work is very important in the different tests.

The Second Year of Technology Education Teacher Training

The second year is devoted to training students in the profession of teaching. This training has three components. The first is a general education for all teachers, independent of subjects or levels, with the aim of acquiring general knowledge and competencies to be a teacher. The second consists of education to become a teacher within a specific subject area, that is, Technology Education or at a specific level, such as secondary school. Students have to deal with knowledge and competencies about teaching, learning and assessing processes. The third is practical training in the classroom: students are placed in charge of some classes throughout the school year for 4–6 hours per week. The student is supervised by an experienced teacher and by a coordinator from the IUFM.

There is a strong articulation between the three parts, organized to support both general and specific aspects of the job. Regarding the transmission of specific knowledge, four points are the focus for this year: team working (in the same subject and with the other subjects), diversity management (diversity of culture, languages, levels, and learning paces), citizenship education and assessment (such as assessing pupils' work, teacher organization and pedagogical projects).

All students who aim to become Technology Education teachers have to dedicate a part of their time to develop a relationship with an enterprise. This activity is based on the widely held idea that Technology Education must relate to organizations. In many cases, students spend four to six weeks in a company where they conduct a project. The theme of this industrial project becomes a report developed into a teaching sequence. It is difficult to find significant meaning in this activity because, on the one hand, the development of the students' enterprise project needs more competencies and time to be successful. On the other hand, the curriculum requires more general knowledge and overview than specific and specialized competencies. Students ask for short periods and a general overview of enterprise, when the enterprise asks for a longer period and a concentration around one specific problem. It is difficult to find enterprises to accommodate all the students. Students have difficulty seeing the relevance of this period to their training as a teacher. Despite these different viewpoints, the simple idea is that the teacher who teaches technology must know and experience the industrial world.

During this year the students get a school appointment as a trainee and they are under the same rules and principles as the state employed teachers; this involves 1,010 TE students who succeed in first year and progress to second year. Usually, a student stays in the same IUFM for the two years. However, as the Ministry of Education is in charge of the students' posting and, in accordance with the accredited number of second year students for each IUFM, a student can move from one IUFM to another. The Ministry can take this decision according to personal criteria for the student, such as their family situation or examination position.

Primary School Teacher Education

Technology Education has a small shared role with science in primary school teacher education. Both years of training must be differentiated similarly to secondary school teacher education.

The first year is fully devoted to preparation for the competitive entry examination, success in which needs a high level of performance but in the various primary school subjects. The main subjects are French (as the mother language), mathematics, sports and educational sciences; these four subjects are compulsory for all students. Students choose two optional subjects in two different groups. In the first group, they can select History and Geography or Sciences and Technology. In the second group, they can choose a foreign or regional language (English, Spanish, Italian or German as foreign languages and Provencal French, Corsican, Basque, Breton, Alsatian as some of the regional languages) or art (musical education and pliable arts). Evidently, the nature of the examination prioritizes academic and theoretical knowledge and students choose the option they studied before their entry at the IUFM. Only about 30% choose the Sciences and Technology option. These students follow an 80 hour course divided in three parts: biology-geology, physics and technology; in addition they follow a 27 hour course in Technology Education.

Those who succeed in the first year enter the second year. The second year is profession-oriented training but according to the non-specialized dimension of the job, students have the same four compulsory subjects (French, mathematics, sports and educational sciences) and they must choose from each group an option they have not studied during the first year; that is, Sciences and Technology is a compulsory subject for those who chose History and Geography in the first year. They have a 50 hour course and for Sciences and Technology, this course is divided in three parts (biology-geology, physics and technology).

At the end, students who have some background in sciences or technology choose this option in first year and approximately 30% of them follow a 27 hour course in Technology Education, but the large majority does not choose this option and follows a 17 hour course in Technology Education during second year. Under these conditions, the place given to sciences and technology education at the primary school level is not really a surprise. Compensating for this lack of training, educational authorities offer a specific in-service course for teachers during their three first years of teaching. The IUFM designs and organizes this 72 hour course over three years.

AN EXAMPLE: IUFM AIX-MARSEILLE

The course presented here takes place in the two years of teacher training. It is based on the industrial project method that takes a very important place in Technology Education in France and, of course, in technology teacher education. Between the introduction of the national curriculum in 1985 and the creation of the IUFM in 1991, this course changed significantly.

General Background

This approach addresses one of the main problems: how to relate production techniques to the human dimensions of technology (Lux & Ray 1970; Lux 1991; Ray 1978; Norman 1998). Different forms of the human dimension are taken into account: production as an answer to human needs, design under-utilization constraints, integration of the solution's cost, the price-quality ratio, and the relationship between product and distribution modes (Amalberti 1996; Bessot & Vérillon 1992). One method emerges as a good integration of these different aspects and allows for designing (or redesigning) a product by incorporating two kinds of functions: that of use (why this object exists) and that of symbol (why I wish to buy this object and not others with the same use).

The entire life of a product, from the initial idea to its recycling, can be described in ten steps (Rak et al. 1992):

- A needs analysis validates the idea in terms of its fit with the needs of the user;
- A feasibility study by which these functions are transformed into technical functions;
- Design is the phase for investigating technical solutions in answer to technical functions;
- Definition is the integration of use, functions, and structure into a global solution that defines the properties of the final product;

- Industrialization studies the organization of the industrial manufacturing of the product;
- Approval describes all the processes to validate solutions in terms of normalization and standardization;
- Production is the phase of manufacturing by which the company makes the products;
- Commercialization describes the product's distribution, advertising and marketing;
- Use concerns all studies about the product's use and maintenance;
- Recycling studies the end of the product's lifetime.

Course Organization

More than the simple addition of different techniques is needed for realizing each operation at each step, whether considered from the reference point or from the teaching point, this method is central to understanding technology teacher education in France.

First Year Course

The first year course is based on the acquisition of the tools, analysis, language and knowledge required to conduct a full industrial project (200 hours). During this course, students have to make their own project that they present as part of the selective entry examination and they have to improve their knowledge, understanding and competencies to realize this approach. With this breakdown of the process that brings a new industrial product onto the market, each step constitutes a collection of specific problems to solve, and the solution to each problem is an input to solving problems at the next step. Problem-solving requires using identified methods and specific techniques like functional and technical specifications, SADT analysis and the APTE method (Jouineau 1986). Synchronization of each step and each task in time presupposes using a project management method (like PERT). The course organization shares the time between theoretical elements, practical applications and practices. Over these classes, students must devote significant personal work time (about 200 hours) to the realization of the project. Table 1 presents the organization of this course:

Table I.
Organization of the Industrial Project Method course

Industrial project method steps	Modules	Description	Duration
1	1	Analysing the needs: competitive market analysis; consumer analysis; resources analysis.	25 h
	2	Functional analysis: use and symbol functions; general prescription	25 h
2	3	Technical analysis: technical functions and constraints definition; technical prescription.	25 h
3–4	4	Designing product: integrated technical solutions; relationship function, structure, form, material; manufacturing prescription; using tool for computer aided design.	25 h
5–6	5	Preparing manufacturing: norms and standards integration; manufacturing organization; workshop arrangement (tools, material, human organization, etc.).	25 h
7	6	Using the tools for mechanical manufacturing: wheel machine, milling machine, computer aided machining.	25 h
	7	Using the tools for electronic manufacturing: computer aided printed circuit designing; computer aided electronic circuit manufacturing; component welding.	25 h
8–9–10	8	Planning the commercialization: determination of the financial aspects of the product; elaboration of a marketing plan; making an advertising campaign; determination of the user book; organization of the maintain net; organization of the recycling process.	25 h

These eight modules cover all stages of the Industrial Project Method. For the students, developing their own project at the same time gives them a strong articulation between theoretical knowledge, competences and the development of practical abilities. They complete their knowledge field through the other subjects, specifically in mechanical engineering, electrical engineering, and business economy and management.

The most important problem to solve in this course is the students heterogeneity based on the previous study they have done. In a different situation, this heterogeneity could present the opportunity for competency collaboration, but in the context of the selective entry examination, there is not really collaboration, and we can observe some competition between students.

Second Year Course

During the second year, after the pressure of the selective entry examination, students have to learn how to organize their teaching, including the different steps of the Industrial Project Method. In the last year of lower secondary school, pupils synthesize all knowledge from the previous three years; this is the main goal of the course (30 hour course over 15 weeks) *Organizing Teaching of the Industrial Project Method.* The course is organized between the university and practice time in a school with pupils. Students try to implement the method they developed during the course directly in class. This organization induces a real dynamic between the pedagogical background, the preparation of the class and the practice with pupils.

One of the obstacles for the young teachers is to provide all the documents necessary to successfully implement the project. To address this problem, students have the opportunity to access distance-resources, specifically through the project Challenge 2000, developed by the IUFM investigation team. Before using this distance-resources system, students devoted the majority of their time to producing documents for pupils, rather than being able to think about the organization of their teaching or its assessment and regulation.

Firstly, students have to analyze the situation in the school, establish a checklist of previous knowledge and determine the required competencies. This provides some perspective to Technology Education in the context and in the school. Then students elaborate on the general organization of their teaching. They start building a pedagogical project according to the different aspects, goals and constraints. Their last job consists of organizing and planning the different sessions for the year. During this phase, student attention is focused on the pedagogical strategy and sequences planning, including the specific competencies according to the National Curriculum. They can use all the distance-resources and, specifically, all the documents they need to organise their teaching. This offer gives them the opportunity to take time to think about class organization, group activities and assessment of their own practice. For that, they have at their disposal the description sheets of each sequence, the development sheets and the annual planning. They have to integrate all the different constraints of their school and the TE teaching conditions. At the end, they have at their disposal a full pedagogical project including the following folios:

- The session development sheets with spatial and time organization, which details the planning of pupils' activities.
- The teacher sheet combines many elements to help teachers in the task: 'everything teachers discuss when they talk about a project but that is rarely written', such as advice, tricks and shrewdness.
- The pupils' sheet details their activities in terms of knowledge, goals, tasks and assessment. This sheet leaves the specific pedagogical method open.
- The resource sheets include documentation, bibliography and references about possible tools, software and machines.

Students can decide the form and the content of each folio and the media used, for example, paper, Internet, Intranet. This course organization is fully articulated with the other courses but also with the practice time they have in the school. The students are more able to concentrate on the global project's elaboration. A large part of the work can be done by distance via Internet and, during the in-class training sessions, teachers can concentrate their time on the different teaching logics, putting these logics in perspective with the subject content as expressed in the curriculum, and with the pupils' learning principles, notably through the question of knowledge construction and teaching organizations on the efficiency of the learning. This collective work allows for a discussion between the teachers about the different choices they make and about the implications of these choices. The final second year student assessment is widely based on this project, the folios they construct and the report about their practice.

TEACHER CERTIFICATION

The first stage of teacher certification is the selective entry examination. Even if the general principles are the same, the nature of the tests is different for primary and secondary school teachers. For primary school teacher certification, students have to take an oral test. The theme can be biology-geology, physics or technology and they comment on a dossier; for Technology Education, the dossier is composed of a technical study of a system and a pedagogical aspect of this system. For the secondary school teacher certification, the examination includes two groups of tests which are exactly the same for all students in France. The first group, taken in January, consists of two written tests. The first test is representative of the technical and scientific knowledge and the second test covers the knowledge about the industrial project method. The second test, taken in June, consists of three oral tests. The first assesses the students' ability to articulate the technical and scientific knowledge with the project knowledge. During the second test, students defend the project they undertook during the training year. The third test assesses the abilities of the students to use the machines, tools and models normally used in a classroom.

Assessment of the second year is organized in two stages. At the first stage, it is the IUFM and at the second stage, it is the Ministry of Education that assesses the candidate. There is some overlap between the two stages, but each is also significantly independent: nobody can be in both juries but the second stage is based on the folio which is elaborated by the first.

The assessment of the second year by IUFM is related to the professional aspects of teaching and school activities, such as team working, project collaboration and social integration. It is based on three points:

- Evaluation of the practice of the trainees in their classes: during their practical period, the trainees are shadowed by at least three people with different roles, all of whom produce an evaluation. The headmaster of the school is consulted about team working, project collaboration, and school integration. The supervising teacher gives an opinion about the classroom practice of the trainee, and about the trainee's ability to integrate into the classroom. The coordinator gives an assessment of the link made by the trainee between the practice and the IUFM courses. This formative assessment assists the development of the trainees and their professional identity.

- The defense of a professional dissertation: developed during the year, under the direction of a supervisor, the aim of this dissertation is to allow the trainees to reflect on their role as a teacher, not in general terms, but from the analysis of a real school situation. One of the goals is for the young trainee to recognize how much work exists about teaching, learning and school, and that it is not necessary to experiment alone to find solutions to some well-known problems. In fact, the structure of the dissertation starts with an initial question, followed by a review of works on this question (or on the field in which the question is integrated). This review must be the basis for some experimentation, and then the data collected analyzed to

improve the initial question. The dissertation is defended in front of a jury composed of three people, including the supervisor.
- The oral test about the different knowledge learned during the training is based on a presentation of a teaching sequence elaborated by the trainee. After a presentation of the sequence (goals, context, organization, pupils, school…) by the trainee, the jury asks questions.

The constituted folio for each student, with all the assessment elements, is transmitted to the employer jury. There is one jury in each region. On the basis of this folio and during a first meeting, the employer jury decides who will succeed. The other students, including all those that were not validated by the IUFM, receive a visit from a member of this jury in their classroom. This person makes a report about the practice of the trainee and their impression after a one hour interview. On the basis of this report, the jury, during the second meeting, decides to employ (to give tenure to) the person (90%), or proposes the person repeats the second year (8%) or excludes them (2%). Noticeably, the main selection was made during the first year but this selection was not made on the ability to teach but on the mastery of the subject knowledge.

CONCLUSION

France was one of the first countries to introduce a TE curriculum for all. Technology teacher education is now well organised in the French context of a selective entry examination. One can observe in the French school a real evolution of practice due to the massive increase of TE teachers. This development comes from the drastic change of student profiles and breaking with the art and craft traditions. This effect is more dramatic since the mid 1990's with the creation of the IUFM and the new standards in TE. TE is now a reality for all pupils within a national curriculum by well-trained teachers. In the same way, the massive development of resources, notably by the IUFM and for example, Challenge 2000 supports and anticipates the subject's evolution. Even so, the place of TE in the French school system is not really stable and is under diverse pressures. The massive increase in numbers of students disrupts the school system and, specifically, the position of the general subjects (such as French or Mathematics) and the pupils who have some problems of knowledge access. Instead of a real and deep reconsideration of the national curriculum in these subjects, the tendency is to devote TE to the treatment of pupils with school difficulties. The redefinition of the national curriculum in terms of TE as a component of general education is one of the ways followed. The contribution of the IUFM, by way of teacher training, has also been important.

Another development has been initiated so that if students do not succeed in the selective teaching entry examination, they do not get a diploma from the IUFM. Many people question this link between the professional qualification and the qualification by a diploma. The idea, common in many countries, is to separate the training from the employment. It is not easy to change because opponents are numerous and with different reasons, like; defence of the state employment status, defence of the French republican school tradition and the difficulty of achieving level 5 of the teachers qualification. But the arguments for change are also numerous, such as the harmonisation of the French teacher training system with the European system, the development of a real vocational training for teachers by giving it more time and the recognition of the high level of competencies required by teachers. The next development will be the re-organization of training with the delivery of a diploma of education at master's level. The question of the place of the selective entry examination is open: it may be successive like now (starting with three study years within the subject, followed by two years of teacher training) or simultaneous as in the majority of countries.

For technology teacher education, this question is important. It is difficult to find students with a profile corresponding to Technology Education. Training students for five years is very different from two years training. In fact, the recognition of five year courses, and the delivery of a diploma like masters and a licence will open the opportunity to develop a PhD in education. These links between Technology Education practices in school, teacher training and investigation in Technology Education are essential and constitute the main challenge for France.

REFERENCES

Amalberti R., (1996). *La conduite des systèmes à risques.* Paris: Presses Universitaires de France.

Bessot A. & Vérillon P., (1992). *Espaces graphiques et graphismes d'espaces.* Grenoble: la Pensée Sauvage.

Charlot B., (1997). *Du rapport au savoir. Éléments pour une théorie.* Paris: Anthropos.

Develay M., (1992). *De l'apprentissage à l'enseignement : pour une épistémologie scolaire.* Paris: ESF.

Ginestié J., (1999). *Techniques scolaires et enseignements technologiques.* Actes des XXIe journées internationales sur la communication, l'éducation et la culture scientifique et industrielle. Paris: éditions du LIREST.

Ginestié J., (2002). The industrial project method in French industry and in French school. *International Journal of Technology and Design Education,* 12(2).

Hill A-M., (1998). Problem Solving in Real-Life Contexts: An Alternative for Design in Technology Education. *International Journal of Technology and Design Education,* 8(3), 203–220.

Jouineau C., (1986). *L'analyse de la valeur: méthode, mise en œuvre et applications.* Paris: Sème édition, ESF - Entreprise Moderne d'Édition.

Lux D. & Ray W., (1970). Toward a Knowledge Base for Practical Arts and Vocational Education, *Theory into Practice,* 9(5), 301–308.

Lux D., (1991). Science, Technology, Society Challenges. *Theory into Practice,* 30(4).

Norman, E., (1998). The Nature of Technology for Design. *International Journal of Technology and Design Education,* 8(1), 67–87.

Rak I., Teixido C., Favier J. & Cazenaud M., (1992). *La démarche de projet industriel,* Paris: Éditions Foucher.

Ray W., (1978). A School-Wide Practical Arts Program: Technology for All Students. *Journal of Epsilon Pi Tau,* 4(1), 4–13.

Technology Teacher Education in Germany

Chapter 4

Gerd Höpken
Flensburg University, Germany

INTRODUCTION

The education system in Germany is funded by both federal and state governments and communities, but it is controlled by the 16 state governments. So each state has a different system of grades and curriculum in education. There are however many similarities in curricula due to federal regulations, working groups of the state ministries of education and national curriculum projects.

Figure 1. German school system.

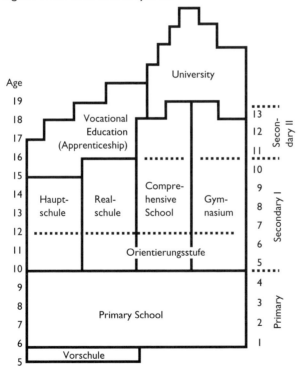

Not all school systems of the German federal states have K–12 or K–13 curriculum frameworks for technology. If there is a subject it is variously called Technik, technisches Werken or Arbeitslehre. Technik is generally organized around a process and systems concept, containing elements of processing material, energy and information as well as the consequences of technology. The main concept of technisches Werken is making, in the sense of working with wood, metal, and plastics.

Arbeitslehre is a subject focusing on the labour process, career, technology, and economics. An example of a German technology syllabus is the Schleswig-Holstein curriculum (Fig. 2[1]).

Figure 2. Schleswig-Holstein Technology Curriculum
(bold: compulsory subject matter).

Socio-technological field of action					
Work and production	**Responsibility of man working with raw material in craftsman like production. Basic course: Communication in technology.** Year 7–9	**Development and employment of machines change place of work and vocation. Interdependence of man and machine in production.** Year 7–9	Industrial production of products for daily use and its impact on conditions of life. Year 8–9		
Transportation and traffic	**Bicycle technology and appropriate use of means of transport.** Year 7	**Car technology and its interactions with man and environment.** Year 9	Technology conceptions for environment, transportation Year 10 project	People develop technology (e.g. air craft engineering) Year 8–10	
Construction-built environment	Former and present ways of constructing bridges. Year 7	People protect and secure themselves –safety systems of yesterday, today, and tomorrow. Year 7–8	**Dwelling in changing times** Year 8–10 project		
Supply and waste management	Wrapping is a burden for environment – recycling relieves environment. Year 7–9	Supplying and disposing garbage of a household under technological, ecological and economic aspects. Year 8–9	**Using energy efficiently and sustainable energies in households.** Year 9–10	Man as consumer – discriminating dealing with the supply of technological products Year 9–10	
Information and communication	Basic electrical circuits and safety education. Basic course: Soldering. Year 7	**Impact of automation technology on man, work, and vocation. From hand control to computers** Year 7–10	Interchange of information, development and impacts. From the drum to wireless telephones. Year 8–10		

[1]Ministerium für Bildung, Wissenschaft, Forschung und Kultur des Landes Schleswig-Holstein (ed.): Lehrplan Technik. Kiel 1997

In recent years the German education system was affected by two major factors, which caused dramatic changes. One is the Organisation for Economic Co-operation and Development (OECD) PISA study which assessed the accomplishments of 14 year old students in 32 selected countries. The other factor is the so-called Bologna process, an agreement of all European Union (EU) nations to change the national system of university education towards a common internationally comparable system that enables students to move between universities in different countries during their studies.

The PISA Study

The PISA study was conducted by OECD researchers. It compares students' accomplishments in 32 countries in three school subjects: reading, mathematics and science. When the results were published in 2001, it was a shock for the German public: (Fig. 3[2]) showing that German pupils ranked in the lower third. One of the main problems for German students was a lack of ability to understand the text of the tasks and the mathematical problems in their context.

Figure 3. PISA Study: Outcome

	Reading	Math	Science		Reading	Math	Science
Finland	546	536	538	Denmark	497	514	481
Canada	534	533	529	Switzerland	494	529	496
New Zealand	529	537	528	Spain	493	476	491
Australia	528	533	528	Czech Rep.	492	498	511
Ireland	527	503	513	Italy	487	457	476
Korea	525	547	552	Germany	484	490	487
UK	523	529	532	Liechtenstein	483	514	461
Japan	522	557	550	Hungary	480	488	496
Sweden	516	510	512	Poland	479	470	483
Iceland	507	514	496	Greece	474	447	460
Austria	507	515	519	Portugal	470	454	443
Belgium	507	520	496	Russia	462	478	460
Norway	505	499	500	Latvia	458	463	459
France	505	517	500	Luxemburg	441	446	422
USA	504	493	499	Mexico	422	387	478
Average	500	500	500	Brazil	396	334	375

---

[2] http://www.agsp.de/UB_Forum/Diskussionsbeitrage/Diskussion_22/hauptteil_diskussion_22.html

One of the consequences of the result of the PISA study was the elaboration and implementation of standards. The Ministers for Education Conference (KMK) held in Bonn in 2002 decided, for the first time, on common education standards throughout Germany. Education standards stipulate what a pupil must master in a given subject by the end of a school year. The Ministers for Education for the federal states reached an agreement on December 4, 2002, regarding the education objectives in the subjects of German, Mathematics and the first foreign language (English or French) for pupils who leave school after the tenth year with intermediate education qualifications. The performance requirements applied as from summer 2004 and are compulsory for all federal states.

With the introduction of education standards, the Ministers for Education for the federal states were reacting to the poor performance by Germany in the PISA international study of schools. They wanted to improve the quality of teaching, raise the performance level and ensure comparability as regards the final examinations in the various federal states[3].

In 2001 a process of elaborating/refining standards for several school subjects was initiated, commencing with the subjects addressed in the PISA study: German language, mathematics, science. These were seen as the core subjects. Technology education (TE) experts tried to bring the strength of technology education into the discussion by promoting the use of language, mathematics and science in a meaningful context. But the states' committees compiled standards only for the so-called core subjects.

In this situation, the Association of German Engineers established its own standards committee with school and university teachers from all over Germany as members. The publication of the International Technology Education Association (ITEA) standards in German[4] was a contribution to this process.

An academic institute of the 16 state governments will develop tests for assessing compliance with the standards from 2006. A sum of EUR 2.5 million has been allocated for a year's work by the institute. Further to this, inter-state comparative tests are planned.

<u>The Bologna Process</u>

In Bologna on 19 June 1999, 29 European Ministers in charge of higher education signed the Declaration for establishing the European Area of

---

[3]http://www.bundesregierung.de/en/artikel-,10001.574260/The-Ministers-for-Education-de.html
[4]Hoepken, G./Osterkamp, S./Reich, G. (ed.): Standards für eine allgemeine technische Bildung. Villingen-Schwenningen 2003 (ISBN 3-7883-0382-4)

higher education by 2010 and promoting the European System of higher education world-wide. In the Bologna Declaration, the Ministers affirmed their intention to:
- adopt a system of easily readable and comparable degrees,
- adopt a system with two main cycles (undergraduate/graduate),
- establish a system of credits, such as the European Credit Transfer System,
- promote mobility by overcoming obstacles,
- promote European co-operation in quality assurance, and
- promote European dimensions in higher education.

Convinced that the establishment of the European area of higher education would require constant support, supervision and adaptation to continuously evolving needs, the Ministers decided to meet again two years later in Prague in order to assess the progress achieved and the new steps to be taken.

The Ministers in charge of higher education of 33 European signatory countries met on 19 May 2001 in Prague to follow up the Bologna Process and to set directions and priorities for the coming years. In the Prague Communiqué the Ministers
- reaffirmed their commitment to the objectives of the Bologna Declaration,
- appreciated the active involvement of the European University Association (EUA) and the National Unions of Students in Europe (ESIB),
- took note of the constructive assistance of the European Commission (EC),
- made comments on the further process with regard to the different objectives of the Bologna Declaration, and
- emphasised as important elements of the European Higher Education Area (EHEA):
  - learning
  - involvement of students
  - enhancing the attractiveness and competitiveness of the European Higher Education Area to other parts of the world (including the aspect of transnational education).

Technology Teacher Education in Germany

The Ministers decided that the next follow-up meeting for the Bologna process should take place in 2003 in Berlin, to review progress and to set directions and priorities for the next stages of the process to the European Higher Education Area[5].

HISTORY

At the beginning of the sixties, there were calls for arts and crafts to include technology. The initial response was an attempt to address the problems young people had when they left school at 14 or 15 years of age when they did not cope with the challenges of a changing working environment. In 1964 the German committee on the education system recommended changes to lower secondary education with the introduction of the subject named Arbeitslehre (work/career education) which included hands-on work.

The second response was a movement starting with college and university teachers of arts and crafts. They questioned the basics of arts and crafts and urged a new orientation aimed towards coping with a technological world. These teachers, together with many other interested people, conducted six conferences on arts and crafts between 1966 and 1977. Since nearly all participants had an arts and crafts background/knowledge-base, it was difficult for them to structure technology for educational purposes.

Figure 4. Approaches in technology education.

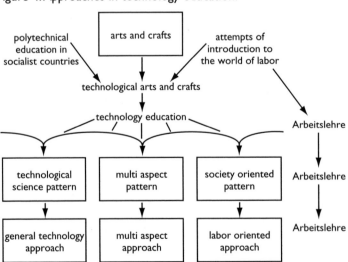

[5]http://www.bologna-berlin2003.de/en/basic/index.htm

Another problem resulted from the Arbeitslehre group who tried to implement curricula through political means. The federal states of Germany are independent in education matters. So, wherever syllabi for technology education were established, their characteristics were dependent on the state government. This resulted in a wide variety of technology/Arbeitslehre curricula. (Fig. 4[6]) Another source of ideas for technology education came from the polytechnic education which was established in East Germany, the German Democratic Republic.

The common problem for all approaches from different sources was identifying the reference point. Engineering, for example is such a widely differentiated field that it cannot be used as a guideline for the general education subject of technology.

Polytechnical education and some of the West German experts in technology education proposed a *general technology* which reduces every technological process to the processing of matter, energy, and information (Fig. 5). For the Arbeitslehre experts, technology can always be reduced to a part of the labor process. For the multi aspect approach, both approaches are part of its theory, but not sufficient to explain technology with all its implications and consequences. An important part of the multi aspect approach is the splitting into *fields of technological action*:

- work and production,
- construction and building,
- supply and waste management,
- traffic and transportation,
- information and communication.

Today, most of the German curricula are structured on the multi aspect approach.

Figure 5. General technology

---

[6]Schmayl, W.: Richtungen der Technikdidaktik. In: Technik im Unterricht (tu) 65 (1992), p. 5 – 15.

OVERVIEW OF TECHNOLOGY TEACHER EDUCATION

Up to the early 1970s, teachers for the primary level (*Grundschule*), the *Hauptschule* and, in part, the *Realschule*, received their professional education at teacher-training colleges (*Pädagogische Hochschulen*), while *Gymnasium* (grammar school) teachers attended university and obtained a degree in a particular discipline, without receiving any special preparation for teaching. This first phase was – and still is – followed by a second, school-based phase of training by means of seminars combined with part-time teaching under the supervision of experienced senior teachers. Since the early 1970s, most teacher-training colleges have been integrated into the universities, and attempts have been made to academically upgrade the courses formerly offered by them, while at the same time introducing some practically and professionally relevant elements into the studies of the future *Gymnasium* teachers. With very few exceptions, nowadays all teachers are educated at universities (first phase, pre service) and teacher education institutions (second phase, in service).

Basically, the integration of technology teacher education for all types of schools into universities has turned out to be hardly more than a formal exercise. Structures and content of TE as well as the typology and characteristics of staff and students of the two major categories of teachers (*Grund-und Hauptschullehrer* – primary and lower secondary school teachers; *Gymnasiallehrer* – grammar school teachers) have remained distinct and separate even inside a university-based education for all. Foundation sciences (in particular educational sciences, psychology and sociology) could not be regarded as the integrating element they were supposed to be, in any of the categories, according to major reform conceptions like the Structure Plan of the German Council on Education[7].

The first phase of teacher education takes at least seven semesters and divides studies into two school subjects, pedagogy and psychology, and ends with a state examination. In the second phase of education the teachers teach ten periods of seminars a week and sit in the lessons of other teachers. After two years of seminars there is a second state examination. One reason for this system is the rigid pay regulations of the German civil service. To belong to the *academic* salary classes, the preparatory studies

---

[7] Sander, Theodor: Structural aspects of teacher education in Germany today - a critical view. In:http://tntee.umu.se/publications/te_structure.html

must have a duration of at least seven semesters. If the studies are shorter, the employees would have a handicap of at least four salary grades. Another reason is that civil servants – and teachers are part of the civil service – have to be qualified in state examinations, which are executed by a commission of the ministry of education, and are not academic exams managed by university staff.

As a consequence of the ongoing discussion on the PISA study, education standards, and the Bologna process, this rigid process of teacher training, teacher employment and teacher payment is being questioned. But so far, no decisions about new ways of teacher certification and teacher employment have been made.

STRUCTURE OF TECHNOLOGY TEACHER EDUCATION

Teacher education in Germany is different to the international system of undergraduate and postgraduate studies. There is a structure of *Grundstudium* (basic studies) and *Hauptstudium* (main studies). Universities have different departments to educate teachers for the three different school types (Hauptschule, Realschule, Gymnasium). Normally one department is responsible for the education of teachers for primary and lower secondary education and another for higher secondary education. There are very few universities which provide technology teacher education for higher secondary education; there are only three federal states with this level of technology curriculum. Recent discussions have suggested aligning the German system of university education to the international bachelor/master system. This will give university students more freedom to switch between teacher education and other courses of study.

Teacher education in the different Federal states of Germany is traditionally well established: primary and secondary teachers get their initial education at the university (since most of the academies and teacher colleges merged in the 1980's[8]). After the university exams there is a two-year period of practical experience in schools, supervised and regularly controlled by the *Studienseminar*, which is organised as the state's responsibility, separately from the universities. They finish the so-called second phase with the final teacher exam. There is a rich variety of service programs run by the states or

---

[8] http://www.comenius.de/projektedetail.cfm?id

state agencies, churches, teachers and independent organisations which offer in-service courses on a volunteer basis (the so-called third phase). Most teachers, but not all, feel a professional obligation to participate in the in-service courses. For the last ten years there have been school development policies in several Federal states to promote school autonomy, accountability and self-evaluation, and to support these developments by school based in-service courses.

Before the outcome of the PISA study was published, the most important push towards innovation came from the results of the Third International Mathematics and Science Study (TIMSS), published in 1997, which provided evidence that the achievement-results of German pupils tended to be about average in international comparisons. The well established teacher education system apparently did not lead to a comparatively higher quality of schooling. Since then many associations, boards, commissions and university advisor groups have analysed the system of teacher education and published recommendations to change. Two of the major recommendations are:

(1) The Expert commission recommended changes in the three levels of teacher education:

a. Teacher Initial Education at the university
- elaborate a core-curriculum in educational science in order to end the arbitrariness
- promote research (classroom research) in the subject studies
- establish centres for teacher education and school research within the universities

b. Induction Phase
- qualify the personnel in the induction phase
- better match between initial and induction phase
- more self-monitoring of beginning teachers in professional learning

c. Professional learning
- establish a system of induction including obligatory in-service course work
- in-service plans at the individual school as a basis of personnel and school development
- elements of competence-based teacher salary in career development

These recommendations were considered as balancing innovation and conservation in the teacher education system, leaving specific initiatives of change to the institutions. The University of Hamburg will be the first to restructure the teacher education according to the recommendations of the Conference of the Ministers of Education and Culture.

(2) The Expert commission in the state of Northrhine-Westfalia (NW) came up with a more critical assessment of the quality of the teacher education system in general. The recommendations are subsequently more radical, based on the opinion that the system cannot be innovated but has to be changed. Even insiders cannot get a sufficiently informed overview on the teacher education policies in the different universities.

The ongoing discussion has a certain relevance to the other Federal states because NW has the largest number of inhabitants with the most universities, and could become a leader to the other Federal states.

The experts predominantly criticised the long duration of teacher-education (5 – 8 years) and the lack of orientation to professional prerequisites. The expert advice is to change aims to:

- strengthen the educational part of teacher education,
- remove the gap between theory and practice by, for example, adding more school experience to the education of natural science teachers,
- strengthen the classroom related qualification of natural science teachers.

The recommendations follow these aims by using different teacher education systems in other European countries as exemplars:

1. Establish more flexibility in the early years of study in order to open different professional options to the students and to help graduates find employment.

2. Consecutive organisation of teacher studies: implement the bachelor exam after six semesters, and after a postgraduate period, the masters degree in education. The curriculum has to be organised in modules which promote subsequent teacher in-service professional development and qualifications.

3. Integration of the second phase into the postgraduate modules.

In general there is significant approval for the recommendations from the education departments of the NW universities, especially related to shortening study times and professionalism in teacher education.

Teacher in Service Development

Traditionally from the university point of view, teacher in-service development is not regarded as an important task. However the proposal to change to consecutive modularization of teacher education will have a significant influence on teachers and school development. The policies of school and teacher development have not yet been implemented. In most Federal states, school development is promoted by the individual school, so quality and evaluation processes must be ensured by implementing a system of school advisors to monitor the change.

AN EXAMPLE: FLENSBURG UNIVERSITY

In accordance with the Bologna process, the European nations are restructuring their university programs. Every German federal state is now in the process of adapting its universities to the Bologna guidelines. Teacher education will consist of six semesters covering two school subjects, and education studies. The model is based on a workload of 1800 hours per year. Students have to accomplish 30 credit points (cp) in one semester, shared among the two subjects and educational studies (pedagogy, philosophy, psychology). One credit point is an equivalent of approximately one hour per semester, depending on the character of the subject.

Figure 6. Structure of studies

final examination		thesis 9 cp	
1. subject 9 cp	2. subject 9 cp	educational studies 9 cp	6.
1. subject 9 cp	2. subject 9 cp	educational studies 9 cp	5.
school practice 3cp			
1. subject 9 cp	2. subject 9 cp	educational studies 9 cp	4.
1. subject 9 cp	2. subject 9 cp	educational studies 9 cp	3.
school practice 3cp			
1. subject 9 cp	2. subject 9 cp	educational studies 9 cp	2.
1. subject 9 cp	2. subject 9 cp	educational studies 9 cp	1.

semester

Bachelor Technology Teacher Training Course of Study

The following structure of technology teacher education courses is that of the Department of Technology Education at Flensburg University (It is yet to be decided whether the degree will be named *Bachelor of Education, Bachelor of Arts,* or *Bachelor of Science*).

1. Objective of the course of studies.

The general objective of the course of studies in technology consists of making the complex technological environment transparent and understandable, as well as enabling it to fulfill the general employment requirements in technically oriented occupations. A goal is therefore the preparation and/or development of factual, action and evaluation competencies in technological fields and is directed toward the following goals:

- Factual orientation in the technological fields of processing matter, energy, and information.
- Introduction to the methods and actions typical of technology in planning, designing, manufacturing, evaluating, using, and waste management.
- Realization of structures and functions of technological systems as well as the conditions and consequences of technology.
- Gaining abilities to consciously participate in decisions on present and future living conditions affected by technology.

According to the multi-dimensionality of technology, three levels of technology which are causal and interactive can be named:

- The factual level covers the acquisition of fundamental knowledge about technological articles and procedures.
- The method level covers the ways of thinking and acting typical for technology, as they arise in invention, innovation and production processes, in accordance with abilities such as creativity, co-operation and communication.
- The value level forms the precondition for opinion and decision-making processes in order to critically evaluate and reflect on the development and use of technology under economical, ecological, ergonomic and social aspects.

It is to be noted that the Flensburg Department of Technology Education represents a general technology approach integrating engineering disciplines which are oriented to a multi perspective understanding of

technology. In accordance with this holistic view of the bachelor studies, technology education encompasses the following areas:
- Material processing in technological systems,
- Processing energy in technological systems,
- Processing information in technological systems,
- Consequences and conditions of technological artifacts and processes.

2. Modules

Module 1: Technology and Manufacturing I. One goal of Module 1 is the introduction to basic content and operational sequences of material processing systems. Technological terms from historical and philosophical views are represented including a discussion of the consequences of technology. The technological content of the first module is focused on wood technology. Processing material and energy are located in the center of the lessons. By using the example of electro-technical contents, the need for electrical energy is brought up for discussion as well as the methods to satisfy this need in a sustainable way. Lessons on the presentation of technology are concerned with teaching and evaluating different presentation methods and techniques. In all technology lessons theoretical and practical-technological components are integrated.

Module I	hrs.	Credit Points
1. Manufacturing technology – theory and application I (wood technology)	4	4
2. Electrical engineering	2	3
3. Presenting technology	2	2

Module 2: Technology and Manufacturing II. A goal of this module is the introduction to substantial operational production processes. The material and energy areas are now supplemented by information, in order to be able to encompass the spectrum of technological processes of transformation. The insights into finishing technologies, from Module I, are extended to manufacturing methods of metal and plastics technology. Beside the actual procedures of material processing, the ecological, economic and social aspects of materials and procedures are discussed.

Under the aspect of communication, drawing and construction procedures are introduced. Exemplars are given to connect specific presentation procedures to the teaching of technology. The super ordinate study goal in this module consists of developing an holistic approach to technology, while at the same time examining the principles of the divisions of labor within technological action processes. Therefore study situations must be created in which drafting, planning, execution and evaluation are holistic, as well as dividing these processes into small performing and/or processing steps.

Module 2	hrs.	Credit Points
1. Graphic design and presentation techniques	3	2
2. Manufacturing technology – theory and application II (metal technology)	3	4
3. Manufacturing technology – theory and application III (plastics technology)	3	3

Module 3: Processing Energy and Information in Technological Systems. A goal of this module of energy and information processing in technological systems is the extension of the system concept, whereby the working together of materials, energy, and information is opened to the solution of technological problems.

In all four sections of this module, a systems approach is used, in which energy and information flows are examined and regulated. Attention is particularly directed toward the interfaces, which make possible the division of systems into subsystems. For all content in this module, application-oriented exercises take place in order to make the behavior of systems accessible. In presenting technological systems there is also emphasis on the application of audiovisual media.

Module 3	hrs.	Credit Points
1. Electronics	2	2
2. Renewable energy	2	2
3. Transporting and storing energy and information	2	3
4. Presenting technological systems using audiovisual media	2	2

Module 4: Machine Systems including Control Processes. On the basis of the system concepts gained in module 3, this module provides further insights into the basic behavior of system components. The content focus is on gearing technology (transmission as interfaces of the power transmission), control technology with electronic elements (analogue and

digital processing of information and definition of appropriate interfaces) and teaching the principles and structures of machine systems.

In this module fundamental transferable capabilities are to be acquired, developed from the linkage of design, manufacturing, use and maintenance, evaluation and optimization of an artifact or a procedure. Technological structures enable the social, humane, economic, ethical and ecological effects to become visible. In these references technology can be seen in its purpose-means relationship and can be placed in its history of development. The content of technology education and teaching are both integrated into the individual module sequences.

Module 4	hrs.	Credit Points
1. Transmission technology - theory and experiment	3	4
2. Control technology using mechanical, electromechanical, and electronic parts	3	3
3. Teaching principles and structures of machine systems	2	2

Module 5: Organizing Manufacturing Processes Considering Consequences Beyond Technology. A goal of this module is gaining knowledge, insights and abilities for solving technological problems related to manufacturing. The new content range includes the topic *human labor* within industrial production together with all the social consequences. While in the first two seminars the labor organization is prominent (first with the observation of material working processes, second in the responsible organization of work in a real casting process), the third seminar deals with automated solutions in these processes. A topic for example is the use of computers in the manufacturing process: for the programming of micro controllers. All the seminars discuss the teaching of the content. In addition, knowledge of technology-didactical concepts and the historical development of the teaching of technology is obtained.

Module 5	hrs.	Credit Points
1. Using and assessing methods of organizing labor	2	2
2. Planning sequences of operations (example: casting technology)	4	4
3. Control technology using micro controllers	3	3

Module 6: Presenting Technology and its Consequences. A goal of this module is the application of the knowledge, insights and abilities from the preceding modules in the solution of concrete problems. The learning processes concentrate here on the methods of the evaluation of technology under functional, economic, social and ecological aspects.

In the first seminar, concrete problems from different contexts, for example the construction, machine, electrical as well as control technology, are solved by technological means. A key skill is the ability to sketch and design technological artifacts as well as to organize and manage technological processes. The presentation of technology focuses even more than before on the consequences of the use of technology. Since the strongest influences these days come from information and communication technology, a seminar of its own was assigned to this field.

Seminars also address and cover the teaching of content. A special emphasis is thereby placed on the field of information and communication within technology education.

Module 6	hrs.	Credit Points
1. Designing, planning, and evaluating technological systems (electrical engineering, machine technology, construction technology, ...)	4	4
2. Presenting technology and consequences of technology in different fields of application	2	2
3. Information and communication technology stimulating technological innovation	2	3

Bachelor's Level Technology with Non-Teachers

The bachelor studies in technology are fundamental for the consecutive masters level studies which lead to the qualification for teaching at different school types and in different school stages. The bachelor certificate in technology can also lead to other employment depending upon the choice of additional modules in a second subject. Using the resources existing at Flensburg University, the following options for other vocational fields are conceivable:

- Consultation and teaching within technology/business concerns, technology/marketing, technology/sales.
- Presentation of technology within technology/museums, technology/fairs, technology/advertisements.

- Technology management within technology/crafts enterprises, technology/smaller production enterprises.
- Maintenance of technology, equipment in educational facilities.

TEACHER CERTIFICATION

Teacher education in the different Federal states of Germany is traditionally well established: primary and secondary teachers get their initial education at the university. After the university exams they teach in schools for two years, regularly supervised by the *Studienseminar*, which is organised by the state rather than the universities. They finish this second phase with a final teacher exam. So there are two levels of certification: first after completion of university studies and second after a period of in-service training. In both cases, the certifying examinations are state examinations, meaning that a board of examiners set up by the ministries of education provides the certification.

No final statement about teacher certification is possible until the federal states' decisions have been made. So at present it is not clear whether future teacher certification will be based on six semesters of study (bachelor level) or eight to nine semesters of study (masters level). A possible solution would be the integration of the in-service part of teacher education into the graduate studies. In this case the teachers would be completely qualified when they leave the universities.

CONCLUSION

There are many challenges facing universities in training technology teachers, such as funding, equipment and closing of university departments. The trend toward a variety of training models will enable teachers to be drawn from a broader spectrum of society. The shortage of teachers will have significant effects and has in some cases resulted in the closure of programs. The main problems to be resolved by the 16 ministries of education include: the decision of whether teachers must have a bachelor or a master degree and who will certify these teachers. Another problem for technology teacher education is the acceptance of technology as a school subject.

An additional problem at the moment is a power struggle between the states and the federal government over influence in the education sector: They will have to compromise, since both parts have to contribute to the funding of this sector.

REFERENCES

Bundesministerium für Bildung und Forschung: (2004). *The Development of National Educational Standards.* An Expertise. Retrieved June 15, 2004, from: http://www.bmbf.de/pub/the_development_of_national_educationel_standards.pdf)

Bundesregierung. (2003) *The Ministers for Education decide on educational standards.* Retrieved June 15, 2004 from: http://www.bundesregierung.de/en/artikel-,10001.574260/The-Ministers-for-Education-de.htm

Hochschul-Rektoren-Konferenz. (2003) *Basic Information: The Bologna Process – Towards the European Higher Education Area.* Retrieved June 15, 2004, from: http://www.bologna-berlin2003.de/en/basic/index.htm

Hoepken, G.; Osterkamp, S.; & Reich, G. (ed.)(2003). *Standards für eine allgemeine technische Bildung.* Villingen-Schwenningen: Neckar Verlag.

Mewe, Fritz and Eberhard, Kurt. (2002) *Ökonomische Anmerkungen zur Pisa-Studie.* Retrieved June 15, 2004, from: http://www.agsp.de/UB_Forum/Diskussionsbeitrage/Diskussion_22/hauptteil_diskussion_22.html

Ministerium für Bildung, Wissenschaft, Forschung und Kultur des Landes Schleswig-Holstein (ed.). (1997) Lehrplan Technik. Kiel: Kultusministerium

Sander, Theodor. (2003) *Structural aspects of teacher education in Germany today - a critical view.* Retrieved June 15, 2004 from: http://tntee.umu.se/publications/te_structure.html

Schmayl, W.: Richtungen der Technikdidaktik. In: Technik im Unterricht (tu) 65 (1992), p. 5–15.

Technology Teacher Education in Hong Kong

Chapter 5

Kenneth S. Volk
The Hong Kong Institute of Education, Hong Kong

INTRODUCTION

As a British colony until the handover to China in 1997, Hong Kong still follows much the same educational system as its former caretaker. Since the handover, changes in government, demographics and economics have played a significant role in increasing the rhetoric on the need for educational change as well as actually enabling some new initiatives to be introduced.

'Technology Education' is a broad classification of subject area in Hong Kong schools, stemming from a recent educational plan that organized subject matter into eight Key Learning Areas (KLA) (Curriculum Development Council, 2000). Each KLA is compulsory for students to study and has broad objectives and benchmarks throughout primary and secondary grades. The Technology Education Key Learning Area (TEKLA) includes the subjects of Computer Studies, Business Studies, Home Economics and Technological Subjects. The latter group would include secondary school subjects such as Design & Technology (D&T), Design Fundamentals, Technology Fundamentals, and Graphical Communications. One inherent weakness of the TEKLA is that the interpretation and implementation of specific KLAs is determined by individual schools' independent management, academic needs, facilities, staff, and students. It is possible then for a school to offer only Computer Studies, so fulfilling the broad objectives of the TEKLA.

For this chapter on technology teacher education in Hong Kong, the subject of Design & Technology as a subset within the Technology Education Key Learning Area is detailed. This subject could be found in the 'academic' schools classified as Grammar Schools, while the technological subjects of Design Fundamentals, Technology Fundamentals, and Graphical Communications are also taught by D&T teachers, but are only found in schools previously classified as Prevocational or Technical. In 1999, Prevocational and Technical schools, which represented around ten percent of the total secondary schools, began to convert their metalworking and technical drawing subjects into new D&T-type subjects under new alternative syllabi. They did this because all Prevocational or Technical schools were required to change their role and mission from being vocational to academic (Education and Manpower Branch, 1997). In reality, much of the content of Design Fundamentals and Technology Fundamentals are included in D&T, so much so that it is now being suggested that the different and somewhat confusing syllabi be reduced into one generic subject called Design & Technology or even Design and Applied Technology Studies.

According to the Curriculum Development Committee (2000, p.3), the objectives of a lower secondary Design & Technology program are to develop students':

- awareness of modern technology and its impact on society;
- understanding in the relationship between technology and other disciplines;
- ability to design with consideration of related design factors;
- ability to explore the characteristics of different kinds of materials and their influences to the environment;
- ability to solve problems logically and creatively, through hands-on and exploratory design and make activities;
- ability to master basic skills in the safe use of materials, tools and machines;
- ability to retrieve, process, present and communicate information and ideas using information technology tools whenever appropriate.

Hong Kong Schools

Currently, the Hong Kong school structure is a six-year primary education, three-year junior secondary, followed by two-year senior secondary with a Certificate of Education examination. A two-year matriculation course leading to the Advanced Level examinations follows Secondary Five and is completed at the end of Secondary Seven. The junior secondary level aims to provide a well-balanced and basic education to all students, whether or not they continue their education past Secondary Three. At the senior secondary levels, the curriculum is more specialized, preparing students for work or university studies. There is a wide range of subjects at this level, but students' educational opportunities at upper levels are largely based on the junior secondary school they were assigned and the specific school's program emphasis and available facilities. Secondary school students would normally focus their studies, depending on school, in either arts or science subjects, with those in the former having little or no opportunity to study D&T.

As an 'international city' with a significant number of residents from throughout the world, about four percent of the schools in Hong Kong are considered International Schools, representing and reflecting the curriculum and culture from countries such as England, United States, Australia, Canada, France, and Japan. The government may subsidize these schools outside direct funding, such as providing land. Approximately five percent of the schools are private, representing religious denominations and commercial enterprises.

Proposals now circulated by the government will change this British-style system to one that more-closely resembles the United States and mainland China, with a three-year junior secondary and three-year upper secondary configuration (Education Commission, 2000). Education will continue to be compulsory for students up to Secondary Three. Although there is no definite timetable, estimates suggest to begin phasing-in changes around 2008, resulting in the need for new curricula, examinations, and the development of standard four-year, instead of three-year university degree programs. Much of the impetus for this change stems from concerns about the overall quality of education, too many school subjects being offered, an over-emphasis on public examinations, and the need for longer university degree studies.

One unique feature of Hong Kong schools is to assign students to a specific 'band' of secondary school. Banding 'tracks' or 'streams' students into different schools based on their ability and academic performance. Schools are divided into three equally proportioned bands, so that Band 1 roughly represents the top 33 percent of students (Education Commission, 2001). This practice has long been a part of the Hong Kong education system and will likely continue for some time.

To assign students to a particular band, a Secondary School Places Allocation (SSPA) system is used. Under this system, internal school assessments in upper Primary grades, scaled by a centrally administered aptitude test, along with the possible school sites available in a child's geographic boundary are the main determinants in the student's placement and academic 'band'. As noted by Volk (2004), Design & Technology programs may vary, depending on the school band, with the type of activity, equipment and even students' opportunity to participate being identified as variables. As an example, lower band schools may emphasize skill development to prepare students for trades.

Design & Technology as a School Subject

Over thirty years ago, it was recommended that Design & Technology should be a component of general education (Hong Kong Education Department, 1974) and facilities were provided in all new schools (Hong Kong Education Department, 1978). Enthusiasm for the new subject was initially high, given government assurance for curriculum and facility support. This enthusiasm soon waned as traditional craft approaches of woodworking and metalworking continued. Furthermore, the subject was never fully introduced in all schools, despite recommendations.

According to the Education Department (2002) statistics, only 298 of the 488 secondary schools (61 percent) currently offer D&T, with most schools only offering the subject up to Secondary 3 (S3) level. With over 250,000 students in Secondary 1–3, very few continue studying the subject beyond this grade as an elective. Certificate of Education D&T examinations at the S5 level were offered in 37 schools, with 495 students taking the examination (Examinations and Assessment Authority, 2004). To further illustrate the relatively minor role of D&T in schools, only 32 students from four schools took the Advanced Level examination.

With the recent push for schools to have more autonomy through School Based Management (Education Commission, 1997), some schools have started to eliminate their D&T programs. Factors such as economics, the lack of student interest at upper secondary school levels, excessive numbers of required subjects, and the continued poor image of the subject are contributing factors to this decision. In this author's visits to over 20 different schools each year for student-teacher observations, D&T programs in schools run from very traditional shops with hand tools used to make craft items such as clocks and key rings with limited design emphasis, to more innovative programs with robotics and authentic design challenges. The interest and motivation of the D&T

teacher, rather than the content suggested in the syllabus, have a lot to do with the type of program offered.

Despite this somewhat depressing state of affairs for Design & Technology, there are some signs the subject's limited popularity and reputation may improve. For one, the aforementioned specific inclusion of Technological Subjects such as D&T as a component in the TEKLA may add some legitimacy and impetus for D&T to be more-widely implemented in schools. Already at the upper secondary level, a new subject in Science & Technology is being introduced for students majoring in art subjects. Features traditionally found in D&T are a significant component of this Science & Technology subject. Furthermore, the inclusion of the TEKLA in primary grades may also start to transform the didactic approach traditionally used to one that includes tools, materials and real problem-solving activities. However, such optimism may be tempered by the reality that most principals and the public will be likely to continue interpreting 'technology' as being 'computers', limiting the potential of a D&T approach and topics in all school levels.

Design & Technology is now available as a subject for females, who only a few years ago were limited to Home Economics. Evidence from studies showing differences in students' attitudes toward technology (Volk & Yip, 1999) and pressure from the Equal Opportunities Commission helped initiate this change. Although there was an immediate benefit of 'legitimizing' the subject for all students in schools with D&T, this was tempered with the reality of increased student numbers contained within limited time and resources. As a result, most schools just doubled the amount of students taking the subject, but with only half the amount of time made available. A common scheduling practice is to have students attend D&T once every six days for only 80 minutes per session. This scheduling of classes tends to trivialize the subject and does not encourage student motivation, continuity of lessons, or the recall of knowledge or skills.

HISTORY

In his first policy address as Chief Executive of the new Special Administrative Region government in 1997, Tung Chee Hwa made the bold commitment to ensure that all new teachers entering the profession would be university graduates with professional training (HKSAR Chief

Executive, 1997). Prior to that time, teachers could be trained through a two-year teacher's certificate in education program, providing they had satisfactory Hong Kong Advanced Level Examination results. Principals could also hire teachers without any professional training, but these secondary school teachers would generally have a bachelor's degree in a relevant discipline. As an example of the professional qualifications at that time, 75 percent of the secondary teachers had professional training, and 67 percent had a bachelor degree or above (Brown, 1997). For technical subjects such as Design & Technology, having relevant work experience or a two-year technical diploma may have sufficed. Estimates were that only 23 percent of the D&T teachers had relevant degrees up to the bachelor degree level (Education Department, 1999).

Around the time of the Chief Executive's policy address, the Hong Kong Institute of Education (HKIEd) opened its new campus. The Institute was formed in 1994 from an amalgamation of the five sub-degree teacher's Colleges of Education located throughout Hong Kong. These colleges were under the auspices of the Education Department and were directly responsible for producing the types and numbers of teachers required. The merger of the colleges and placing HKIEd under the University Grants Committee enabled new bachelor degree programs to be introduced, taking advantage of new staff, facilities and government encouragement. Other universities such as the Hong Kong University and Chinese University also prepared teachers, but their emphasis was on 'academic' secondary subjects, with HKIEd focusing on primary school teaching as well as being the sole provider for 'cultural subjects' which included D&T and Home Economics.

Despite calls for an all-graduate all-trained profession and the momentum given the new Institute, reality and economics soon prevailed. It was optimistically felt that those previously-certified teachers who did not hold degrees would be required to upgrade and/or new degree holders would need professional training before teaching. This did not come to pass. In essence, the government softened their position by not placing a timetable on implementation. Furthermore, any encouragement for existing teachers to upgrade their skills was not matched by incentives such as tuition assistance, release time to attend class, promotion, or pay raises. School principals were still free to hire new degree holders without professional training. With the unemployment rate nearing eight percent, and the resulting drastic fall in starting salaries for graduates such as engineers being approximately half of those for teachers, teaching as a profession became attractive to those seeking more lucrative and secure careers. While encouraging all people entering teaching to obtain professional training, it was still not required.

One other event that had some impact on Design & Technology was the change in the government structure and management of education. The traditional structure included the Education Department having control over secondary and primary schools through the curriculum, the qualification of teachers, and the supply of new teachers. In regard to the curriculum, the Curriculum Development Council designed the syllabus, the Curriculum Development Institute produced curriculum material, and the Examinations Authority as a quasi-independent organization produced the Certificate of Education and Advanced Level examinations. For tertiary education, the University Grants Committee generally gives the eight universities in Hong Kong complete freedom with their policies, curriculum, staffing and admissions policies. All organizations were under the umbrella of the Education and Manpower Bureau, with a civil servant, non-educator as its head.

In 2002, responding to criticism over accountability and the deteriorating government budget situation, the Chief Executive instituted a 'ministerial' system to better coordinate different bureaus. As a result, Professor Arthur K.C. Li, a surgeon by trade, was appointed Secretary for Education and Manpower in August 2002. This reorganization immediately began to affect all educational sectors, with tertiary institutes being scrutinized about overlapping degree programs, their governance, and lack of collaboration. This later concern would eventually impact the D&T teacher preparation program being offered at HKIEd.

OVERVIEW OF TECHNOLOGY TEACHER EDUCATION

As described earlier, the traditional Design & Technology teacher was prepared through a two-year Certificate of Education program. This training was provided at the Technical Teachers' College located in the Wanchai district of Hong Kong. Through this program, S7 graduates with technical subject exam qualifications were admitted. The technical laboratories at the College, up until the merger into The Hong Kong Institute of Education and opening of the new campus in 1997, would be considered minimal, at best. As the College was under the management of the Education Department, it produced teachers that mimicked the technical and pedagogical practices followed in secondary school programs. The focus was on skill-development using hand tools. The College had small woodworking, electricity and metalworking labs. Equipment such as CNC and CAD were not included, as schools did not have such topics in the syllabus.

New students entering the teacher preparation program were first given a practical admissions test that required two days to complete. It was largely hands-on and would typically require the construction of a wood box with hand-cut dovetails, and then the construction of a sheet metal box with a hinged top. Needless-to-say, this examination severely limited the number and type of students entering the program, not to mention perpetuating the craft tradition.

With the formation of the Institute in 1994, the recruitment of new D&T lecturers, the new campus being designed, and the control of teacher preparation programs shifting from the Education Department to the University Grants Committee, innovations were put in place. These included changes to the existing Certificate in Education program, the recruitment of traditionally unrepresented students into the program, and the design of new facilities that would eventually enable even more significant changes to occur.

The method of recruiting new D&T teachers into the program received early attention. Specifically, the skills-test for admission largely determined that all potential students had been through a senior secondary D&T program. It also precluded any females from attending, since they had not had the opportunity to participate in the subject and so gain the skills. In order to address this problem, an admissions test was developed which more-resembled a recruitment tool, to show students what was done in the HKIEd program, and not necessarily what was being done in secondary schools. For instance, the exam required students to construct a simple electric circuit using breadboards, as well as to design a structure for an 'egg-drop' challenge. Entrance to the program was not predicated on whether their egg survived uncracked, but rather by a short essay on 'what technology is' and their reasons for wanting to be a teacher. The result of this new approach was immediate, with many male students now having talents and academic abilities outside a traditional D&T experience and approximately 30 percent of the students being female. In essence, the seeds were being sown for future changes in the school curriculum.

The facilities for Design & Technology at the new campus gave flexibility and freedom to expand the curriculum. As the new campus was originally planned by the Education Department before the Institute's current staff was recruited, maintaining tradition and status quo was evident. For example, the five workshops originally proposed included woodworking, metalworking, plastics, electricity and drafting – the subject content followed in schools. So unimaginative were the planned D&T facilities, that the drafting room included only board drafting; as computers were not then part of the D&T secondary school syllabus.

The new lecturing staff quickly proposed a new program arrangement and facilities to reflect a philosophy based on subject integration, rather than segregation; new technology, rather than old; and an exploration of design and creativity, rather than a simple mastery of skills. Some of the changes included a 3-D Studio, combining wood, plastics and metals; an Exploring Technology lab which introduced a variety of 'modular' activities; a Control Technology lab which also included electronics; a CAD (computer-aided design) lab with a variety of software; and finally a Manufacturing lab with flexible manufacturing and robotics capabilities. Obviously, the resulting facilities enabled new programs to be developed, but also served as a model for changes that were to later take place in secondary schools.

STRUCTURE OF TECHNOLOGY TEACHER EDUCATION

The Hong Kong Institute of Education is the sole provider of Hong Kong's Design & Technology teachers. As the Institute's sole mission is to prepare teachers and school administrators, there are programs for kindergarten, primary and selected secondary school subjects. Secondary school subjects taught at HKIEd are classified under the category of 'cultural subjects', which includes Art, Music, Physical Education, Home Economics, Business Studies, and Design & Technology. Large numbers of English and Chinese language teachers are also prepared at HKIEd.

The degree programs offered at HKIEd are a four-year, full-time Bachelor of Education, Secondary (FT BEd(S)) and a three-year, part-time

Bachelor of Education, Secondary, Mixed Mode (MM BEd(S)). The MM BEd(S) is for those practicing teachers who studied the old Certificate in Education program and wanted to obtain a degree. The FT BEd(S) classes are held during the weekdays, while the MM BEd(S) classes are held in the evening, Saturdays and during the summer.

The Institute also offers Post-Graduate Degree in Education (PGDE) programs in Design & Technology. These are run in either a one-year, full-time mode; or in a two-year, part time mode for those practicing teachers without any professional training who have obtained teaching positions. As D&T teachers or teachers of technical subjects such as Electronics could come through an academic degree route, the PGDE programs were developed as an alternative mode for their professional development.

Both the BEd(S) and PGDE programs required extensive validation exercises. These exercises involved both internal and external validations, with the latter requiring a panel of outside experts in subject matter and curriculum, as well as representatives from Hong Kong schools. An enormous amount of staff time was required to not only plan the programs, but to prepare the very detailed and thorough documentation on the Institute, its lecturers, the program structure, and details on each class (module) offered. The whole validation process was managed by the Hong Kong Council for Academic Accreditation (HKCAA), an independent statutory body established by the Government to give advice on the academic standards of post secondary programs in higher education institutions. Since 1990, the HKCAA has been commissioned by the Government to conduct academic accreditation for non-university institutions such as HKIEd and their Bachelor degree programs. In 2004, the Institute received self-accrediting status and so extensive reviews for program approval or major revisions are no longer necessary.

The Bachelor of Education in Design & Technology

This section describes the overall program structure for the current Bachelor of Education (Secondary) in Design & Technology, validated in 2000 and revalidated in 2004. The revalidated FT BEd(S) is a 136-credit point program, while the MM BEd(S) requires 66-credit points. Each credit point is equivalent to 10 hours of classroom instruction. For the subjects of Art, Music, Physical Education, Home Economics, Business Studies and Design & Technology, the total student intake each year (governed by UGC) would be approximately 100 for the FT BEd(S) and 80 for the MM BEd(S). The anticipated class size for each subject would be approximately 15 students. Academic majors would be in the above subjects, with minors offered in Civic Education, Inclusive Education, Information Technology, Mathematics, Putonghua (Mandarin) and Science.

The prime focus of the BEd(S) program is for students to acquire academic knowledge and experience in discipline studies, coupled with effective teaching strategies in these subjects (HKIEd, 2003). It is based on a philosophy of preparing competent teachers who are capable of teaching at all levels of secondary school and "have undergone a rigorous programme of academic, professional and general studies and developed the intellectual stature and critical thinking skills expected of a university graduate" (p. 9).

As stated in the Submission for Programme Revalidation (HKIEd, 2003, p. 9), BEd(S) graduates are expected to:

- Display academic competence in their chosen field with subject knowledge levels appropriate for teaching through senior secondary for their major subject and junior secondary for a minor subject;
- Demonstrate commitment and competence in assisting pupils, regardless of their learning style and needs, gender, ability and socio-economic background;
- Demonstrate appropriate levels of subject knowledge, pedagogical knowledge and skills, both in theory and practice;
- Demonstrate appropriate levels of proficiency in relevant technology, and knowledge of other subjects to equip themselves as broadly competent facilitators of learning;
- Be able to foster in themselves and, in turn, their pupils the ability to appreciate the significance of moral and social values, and the importance of intellectual, creative, aesthetic, affective, and physical development;
- Display positive attitudes and competence in working collaboratively with pupils, colleagues, school administrators, and other professionals, and foster positive relationships between school, home, and community;

- Show understanding of the social, political, economic, and cultural influences upon education;
- Display bilingual proficiency in English and Chinese for academic and professional purposes; and
- Display positive attitudes toward continuous self-learning and professional development.

The BEd(S) program contains five domains: Academic Studies, Professional Studies, General Education, Field Experience, and the final year Project. Table 1 details the credit point (cp) allocation for both the full-time and mixed mode Bachelor of Education (Secondary) programs.

Table 1.
Full-Time and Mixed Mode BEd(S) Credit Point Requirements.

Domain	FT BEd(S)	MM BEd(S)
Academic Studies		
Major Discipline	48	33
Major Methods and Pedagogy	12	6
Minor Discipline	15	
Minor Methods and Pedagogy	3	
Professional Studies	24	9
General Education	12	6
Field Experience	12	6
Project	6	6
Total	136	66

The Academic Studies domain consists of 78 cp in the FT BEd(S), and 39 in the MM BEd(S) for teachers with two-year Certificates in Education. The academic major consists of 48cp of discipline-based studies and 12cp of methods and pedagogy. The academic minor requires 15cp of discipline-based studies and 3cp of methods and pedagogy. For the MM students, there are 33cp for the major discipline studies and 6cp for the methods. MM BEd(S) as these students do not undertake studies in a minor area. Given that the Institute has as its only mission the preparation of teachers, for classes in the subject disciplines, some time may be devoted to ways of applying their skills and knowledge in school situations.

The Professional Studies domain is extensive and contains 24cp of study in the FT BEd(S) and 9 cp for the MM BEd(S). The eight modules included focus on areas of human growth and development, school curriculum and assessment, philosophy of education, classrooms as safe and caring environments, inclusive education, and developing a critical capacity as future teachers. These topics are seen as essential competencies that beginning teachers must possess.

General Education comprises 12cp in the FT BEd(S) and 6cp in the MMBEd(S). Classes in this area are outside the Academic or Professional Studies domains. One class on Hong Kong Studies which provides an overview of Hong Kong and its interaction with China is required for FT BEd(S) students.

Field Experience enables the gradual induction into the profession and consists of 16cp for the FT BEd(S). Teachers in the MM BEd(S) still have a 6cp field experience component, but this is during their actual teaching time and largely involves reflective journals and visits from lecturers. First year FT BEd(S) students participate in organized visits to help them focus on educational contexts and practice. In the second year, students visit schools one day a week to observe and reflect on lessons. In the third year, they have a four-week block practice experience to begin teaching in their academic major area. During the last year, students undertake a 12-week block practice, to teach in both their major and minor areas.

The last component in the BEd(S) program is the final year Project. This 6cp 'independent study' is on a selected topic of relevance to the student. Students work closely with lecturers who provide advice on topics and methodology. Generally, a 4,000 word paper (or equivalent) and final seminar presentation is required.

Table 2 provides the curriculum framework for the four-year, full-time BEd(S) program.

Table 2.
Four-Year Full-Time Bachelor of Education (Secondary) Program.

Year	Semester	Taught Classes (120cp)					Field Experience	cp
1	1	AS Major 9cp	Prof. Studies 3cp	General Education 6cp				18
1	2	AS Major 9cp	AS Minor 3cp	Prof. Studies 3cp	General Education 3cp		Education Visits	18
2	1	AS Major 9cp	AS Major Methods 3cp	AS Minor 3cp	Prof. Studies 3cp		One-day-a-week School Attachment	18
2	2	AS Major 9cp	AS Major Methods 3cp	AS Minor 3cp	Prof. Studies 3cp			18
3	1	AS Major 6cp	AS Major Methods 3cp	AS Major 3cp	Prof. Studies 3cp		Block Practice 4 weeks	15
3	2	AS Major 6cp	AS Major Methods 3cp	AS Major 3cp	AS Minor Methods 3cp	Prof. Studies 3cp		18
4	1	Block Practice 12 weeks						18
4	2	Project 6cp	Prof. Studies 3cp	General Education 6cp				15

The structure requires students to progress together through the program, taking the same classes. For example, in Year 1, Semester 1, all students would take Human Development as their professional studies class. This progression also holds true for the Academic domain, with students staying together throughout the program. The only time D&T students from other programs may be together is when classes are scheduled for both full-time and part-time groups. To facilitate this, all classed are fixed at 3cp, and the program structured, as much as possible, to have classes offered during the same semester. This arrangement provides more flexibility and economy for potentially small groups of students, and has the added benefit of bringing full-time and part-time students together to share their experiences, ideas and energies.

The Post Graduate Degree in Education for Design & Technology

The Post Graduate Degree in Education (PGDE) program for Design & Technology is offered for university graduates with a recognized degree. Pertinent degrees might include for example, architecture, engineering, or graphic arts. The program contains four domains: Major Subject Studies, Professional Studies, Elective Studies and Field Experience.

The Major Subject component carries nine credit points. It covers the professional features relevant to the specific subject. Professional Studies contains four three credit point classes in generic educational foundation topics. Elective Studies contains six credits and enable students to select topics of an interdisciplinary nature. Classes such as Multimedia Design and Teaching through the Integration of the Arts would be popular electives for D&T PGDE students. Finally, Field Experience provides eight credit units of structured supervision. Students in the one-year program are placed in schools with trained supporting teachers, while part-time (practicing) teachers are matched with a colleague in their school who provides observational feedback and support.

Although the aims of the program largely parallel those contained in the BEd(S), given the academic degree already obtained by PGDE students and the elevated reputation associated with such degrees, it is expected that principals would prefer PGDE graduates to teach senior secondary D&T over BEd(S) graduates.

AN EXAMPLE: Hong Kong Institute of Education

The overall aims of the BEd(S) in Design & Technology at The Hong Kong Institute of Education (HKEd, 2003) are to equip future teachers to be:

- Technologically competent in a variety of areas relating to the teaching of D&T;
- Well versed in design theory and processes to identify and solve problems;
- Proficient in planning, implementing, and evaluating D&T units of instruction;

- Able to identify and select content, develop practical/laboratory experiences, and safely manage facilities in order to meet the diverse needs of students;
- Cognizant of the need to pursue lifelong learning in order to further develop their technological, design and pedagogical skills.

To achieve the aims, students complete classes organized around four main strands:

- Historical, Contextual, and Theoretical Aspects of the Discipline
  - *Graphical Communication* and *Design History and Theory* introduce students to theory and principles relating to idea generation and design development;
  - *Exploring Technology* provides students with general framework for the systems and organization of technology;
  - *Technology and Society* covers technology's historical, social, and ethical impacts.
- Practical Skills, Exploration, Application and Discipline Studies
  - *Material Processing* and *Manufacturing Processes* introduce knowledge and skills on tools, materials and processes commonly used in D&T settings;
  - *Electricity and Electronics, Structures,* and *Mechanisms* provide students with breadth of experiences and skills;
  - *Computer-aided Design, Control Technology, Information Technology in D&T, Power & Energy,* and *Digital Electronics* build on the earlier experiences.
- Pedagogical, Professional Knowledge and Competencies
  - *Design & Technology in the Secondary Curriculum and Classroom* and *Advanced Secondary Methods of Teaching Design & Technology* provide introductory specialist training in the methods and approaches to teach D&T;
  - Inquiry and Critical Issues classes serve to focus participants' understanding of their role as a teacher.
- Foundation for Postgraduate Studies
  - The *Project* is the culmination of all three strands and is undertaken in the final semester.

Table 3 provides a list of the classes included and the sequence in which they are taken.

Table 3.
Design & Technology Classes in the BEd(S) Program.

Strand	Mode	Title	Credit Point	Year (Semester) FT
Historical, Contextual, and Theoretical Aspects of the Discipline	FT only	Graphical Communication	3	1(1)
		Exploring Technology	3	1(2)
	Offered in both FT & MM	Design History & Theory	3	1(1)
		Technology and Society	3	1(2)
Practical Skills, Exploration, Application and Discipline Studies	FT only	Material Processing	3	1(1)
		Manufacturing Processes	3	1(2)
		Electricity and Electronics	3	2(1)
	Offered in both FT & MM	Structures	3	2(1)
		Computer-aided Design	3	2(1)
		Control Technology	3	2(2)
		Mechanisms	3	2(2)
		Information Technology in Design & Technology	3	2(2)
		Product Design and Development	3	3(1)
		Digital Electronics	3	3(1)
		Computer-aided Manufacturing	3	3(2)
		Power & Energy	3	3(2)
Pedagogical, Professional Knowledge and Competencies	FT only	Inquiry into the Teaching and Learning of D&T	3	2(1)
		D&T in the Secondary Curriculum & Classroom	3	2(2)
	Offered in both FT & MM	Advanced Secondary Methods for Teaching D&T	3	3(1)
		Current Issues in Design & Technology Education	3	3(2)
Foundation for Postgraduate Studies	Offered in both FT & MM	Project	3	4(2)

With the variety of equipment found in the laboratories, students are expected to spend a considerable amount of time outside class on assignments. Only the Design, History & Theory, and Technology and Society classes do not utilize the labs. Typical projects within a class would combine theory and practice, but in a practical way that relates to the teaching of D&T in schools. For example, in the Mechanisms class, students might build simple robots, while in Digital Electronics, activities suitable for secondary schools might be investigated and/or designed.

TEACHER CERTIFICATION

The Education and Manpower Bureau requires that any person who teaches in a school must be registered as either a Registered Teacher or Permitted Teacher (Education and Manpower Branch 2004). Registered teachers are persons who possess the approved teaching qualifications and/or experiences laid down in the Education Ordinance. Graduates from the Hong Kong Institute of Education meet this requirement.

Permitted teachers do not have to fulfill such professional training requirements, but are given a permit to teach only specified subjects in specified schools. It is estimated that ten percent of Hong Kong teachers remain untrained, with higher percentages found in technical subjects such as D&T (Education and Manpower Branch, 2004). The first step a permitted teacher must take is to find employment in a school, so that the school supervisor can file the necessary papers on their behalf. Although the minimum academic qualification stated to be a permitted teacher is five HKCEE passes, normally a university degree or at least a higher diploma is required.

As stated earlier, the Government's long-term policy will require all new teachers to be professionally-trained and degree holders. With the two-year Certificate of Education replaced by four-year BEd programs, from the 2004/05 school year all pre-service primary and secondary teacher education graduates from HKIEd, will have BEd degrees.

Tragically, graduates from other university programs may also be degree holders, but not professionally trained.

CONCLUSION

The future direction for Hong Kong's Design & Technology Teacher Education is uncertain. Policy initiatives, new bureaucratic structures, and severe limitations in resources assure one thing is certain – change is inevitable.

Given the inclusion of D&T as part of the umbrella Technology Education Key Learning Area, there could be an explosive need for new teachers in the 40 percent of schools that do not currently offer D&T-type programs. Should the TEKLA actually change the way the subjects within the learning area are structured, such as in an interdisciplinary manner, then the skills and knowledge from D&T teachers could be highly desired and sought. However, skepticism remains due to the marginal importance and negative reputation of the subject.

As for the preparation of D&T teachers, it is quite possible the current BEd(S) program at HKIEd will change into a different model used to prepare teachers in mathematics and science. In collaboration with the University of Science and Technology, which students attend for four years and receive all their academic coursework, they attend HKIEd for one semester each of the last two years for their professional training and experience. The government's desire to increase collaborative efforts for economic and efficiency justifications could drive this change. Should this be the case, the few remaining labs and lecturing staff at HKIEd would only be used for professional studies purposes.

Like Technology Education programs elsewhere throughout the world, Design & Technology in Hong Kong faces many challenges. It remains to be seen whether the recent changes to the teacher preparation program and the resulting introduction of new, talented and technically proficient teachers will assure a viable and vibrant future.

REFERENCES

Brown, H. (1997). *Teachers and training.* In G. Postiglione and W. O. Lee (Eds.). Schooling in Hong Kong. p. 95-116. Hong Kong: Hong Kong University Press.

Curriculum Development Committee. (2000). *Syllabus for design and technology: Forms I-III.* Hong Kong: Government Printer.

Curriculum Development Council. (2000). *Learning to learn: The way forward in curriculum development.* Hong Kong Special Administrative Region: Author.

Education and Manpower Branch. (1997). *Development of the new technical curriculum for prevocational and secondary technical schools.* Retrieved June 9, 2003, from: http://cd/emb.gov.hk/tech/tech_sub/curriculum.html

Education and Manpower Branch. (2004). *Teacher registration.* Retrieved May 12, 2004, from: http://www.emb.gov.hk

Education Commission. (1997). *Education Commission report no. 7: Quality school education.* Hong Kong: Government Printer.

Education Commission. (2000). *Learning for life, Learning through life. Reform proposals for the education system in Hong Kong.* Hong Kong: Government Printer.

Education Commission. (2001). *Secondary one places allocation system: Rationale for reform of secondary one places allocation system.* Hong Kong: Central Government Office.

Education Department. (1974). White paper: *Secondary education in Hong Kong over the next decade,* Hong Kong: Government Printer.

Education Department. (1978). *White paper: The development of senior secondary and tertiary education,* Hong Kong: Government Printer.

Education Department. (1999). *Teacher survey 1998.* Hong Kong: Government Printer.

Education Department. (2002). *Teacher survey of technology subjects.* Hong Kong SAR: Government Printer.

Examinations and Assessment Authority. (2004). *Certificate of education examinations.* Retrieved May 23, 2004, from: http://eant01.hkeaa.edu.hk/hkea/

HKSAR Chief Executive. (1997). *Chief Executive's policy address: Building Hong Kong for a new era.* Retrieved 19 March, 2004, from: http://www.policyaddress.gov.hk/

Hong Kong Institute of Education. (2003). *Bachelor of education (honours) (secondary) programme: Submission for programme revalidation, Part II.* Hong Kong: Author.

International Technology Education Association. (2000). *Standards for technological literacy: Content for the study of technology.* Reston, VA: Author.

Volk, K.S. & Yip, W. M. (1999). Gender and technology in Hong Kong: A study of pupils' attitudes toward technology. *International Journal of Technology and Design Education.* 9, 57-71. Dordrecht, The Netherlands: Kluwer Academic Publishers.

Volk, K.S. (2004). Academic banding and pupils' attitudes toward technology: A study of Hong Kong's selective school structure and design & technology programs. *International Journal of Vocational Education and Training* 12(2). Knoxville, TN: International Vocational Education and Training Association.

Technology Education in Israel

Chapter 6

Moshe Barak
Ben-Gurion University of the Negev, Israel
Arley Tamir
Science & Technology Education Systems Consultant, Israel

INTRODUCTION

Israeli society is in the midst of multiple changes having consequences that will increasingly affect the day-to-day lives of its citizens. The ideological perceptions that had guided the founding leaders of our State and their followers for over three decades are disappearing and new perceptions are taking their place. From a country that was shaped, to a great extent, according to the ideologies of equality and fraternity having the aspiration to build a new kind of society, Israel is today becoming a country adopting the values of capitalism, competitiveness, free market and globalization. Among the other expressions of this change are the reduction in public expenses and the privatization of welfare services, health services, employment and parts of the education system.

During the last decades of the 20th century, Israeli industry underwent a dramatic change. At the beginning of this millennium, more than one third of Israel's industrial exports were in the areas of communication, electronics, software, information technology, Internet, sophisticated medical equipment and advanced agro-technology. While traditional manufacturing industries such as metal, food and textiles were gradually diminishing, many international companies such as Intel, IBM, Microsoft and Hewlett-Packard expanded their research and development centers in Israel. Another change that took place over the past decade was the transition to a 'knowledge society' (Drucker, 1994). Israel is today among the greatest users (per capita) of computers and the Internet in the workplace, at school and at home. Consequently, many people identify technology with electronics, computers and communication systems, and expect that technology education will center on these areas. Another aspect of the major role of the high-tech companies in the Israeli economy and society is the growing involvement of many of these companies in projects aimed at fostering science and technology education for young children.

However, leading figures from sophisticated industries often talk about science and technology education, but refer specifically to mathematics, science and computers. Thus, clarifying the objectives of technology education and its contribution to preparing high school graduates for a successful integration in today's complex and dynamic society, as knowledgeable users and beneficiaries of future technologies, is critical for reducing the gap that exists between the image of technology and engineering in the eyes of the general public and decision-makers, and the practical status of technology studies in K–12 education.

The rapid socio-economic changes are breaking down old social orders. The socio-economic shake-up is placing questions on today's agenda regarding not only the cost of public education but also the most suitable and required objectives of education for the coming decades. The role the country is playing in ensuring that an appropriate education is provided to all population groups and in reducing poverty, and the ways scientific and technological excellence is being fostered among the young generation are some of the main issues being tackled.

The present technology studies in Israel are spread throughout the K–12 education system. Over the past decade, new study programs have been developed for elementary and junior high schools that are incorporating science and technology studies into a combined framework based on the Science-Technology-Society (STS) approach (Yager, 1996). The current program for junior high schools, for example, includes the following topics: Materials, Structures, Properties and Processes; Energy and Interaction; Technological Systems and Products; Information and Communication; Earth and the Universe; Organisms; and Ecosystems. This program underlines the development of pupils' thinking and learning skills as a central objective of science and technology studies, for example, "logical thinking, critical thinking, reflective thinking, lateral and vertical thinking, skills for data collections, processing and representation, scientific inquiry, design, and problem-solving" (Ministry of Education, 1996).

A fairly developed system for technology studies programs based on engineering subjects such as electronics, electricity, mechanics and computer sciences is being taught in senior high schools. In Israel these technology programs are part of the matriculation studies and exams, and offer partial recognition for acceptance to the country's universities. However, these studies are also being taken at a practical level by pupils who do not expect to attend university immediately after completing their high school studies, but may pursue higher studies in a technical college or take advantage of employment opportunities.

Over the past decade, the country has invested significant resources in establishing science and technology centers located next to schools in large cities and peripheral areas with the aim of developing a new curriculum and introducing computers and communication systems into the schools. Simultaneously, interdisciplinary science and technology programs for teachers were initiated and implemented. Comparative international surveys showed that Israel is rated high among developed countries in terms of its allocation of resources to education, relative to its gross national product. In spite of this, combined with the fact that computers and Internet in mathematics and science studies are used very extensively in schools, a decrease in scholastic achievements by Israeli pupils was found in national surveys and in comparative international research studies. For example, in the international mathematics test administered in 1999 by Trends in International Mathematics and Science Study (TIMSS, 1999), Israel was ranked 28th in mathematics and 26th in science out of 38 countries participating in this study. While, the achievements of Israeli pupils in the 1964 test were among the highest of the Western countries, Israel was ranked behind several third-world countries in the 1999 test.

Following mounting criticism about the unsatisfactory achievements of the education system, the government of Israel established the National Task Force for the Advancement of Israel's Education (NTAIE) in 2003 aimed at reviewing the existing system and suggesting a suitable comprehensive reform. One year later, the NTAIE presented their conceptual report and suggestions to the government (NTAIE, 2004), including changes in the structure of the K–12 education system, the main concept of its curriculum, and pre-service and in-service teacher training frameworks. After the government accepted the NTAIE report and made plans for its implementation, technology education seemed to be undergoing, as are other parts of the educational system, a period of public examination regarding its necessity, methods and achievements.

As highlighted earlier, the present technology studies in elementary and junior high schools appear as components in the integrated 'science and technology' program, which is compulsory for all children. Beyond the difficulties in implementing this curriculum (either in the scientific or technological part), the present concept of the program is quite acceptable and no major revisions are planned for the near future. In high schools, the present technology studies are elective and are taken by pupils with a wide range of achievements and motivation. In relation to the high-achieving pupils, the current direction is to emphasize the engineering-scientific-mathematical outlook of technology studies. For instance, most of the pupils taking high-level courses in technology are concurrently taking a course in physics, at the same high level.

The future of technology studies in high schools for the lower achieving pupils is also under debate. One of the central issues bothering educators and decision-makers in Israel is the increasing socio-economic gap between different groups in Israeli society. A significant indicator of this gap is the under-representation of pupils from low-income neighborhoods in higher education, which undermines economic growth and increases income inequality (ASCFA, 2000; Coles, 1999; Forsyth and Furlong, 2003). Consequently, scholars and decision-makers have raised questions about the role of technology education in increasing the scholastic achievements of pupils from disadvantaged areas and in fostering their aspirations and confidence to pursue higher studies, essentially in engineering-related areas. Partial answers to these questions have been given in several studies that explored the potential of technology studies, apart from enriching pupils' conceptual understanding and knowledge, by cultivating pupils' learning skills and raising their motivation to learn (Barak, 1994, 2004; Doppelt and Armon, 1999).

HISTORY

The roots of technology education in Israel may be found in agriculture and craft workshops in elementary schools and in vocational education in high schools, as early as the first quarter of the 20th century. In the early 1970s, the education system was changed to three educational sectors: elementary school (Grades 1–6), junior high school (Grades 7–9) and senior high school (Grades 10–12). This structural reform also involved a change in the declared objectives of technology education, as seen in later formal documents (Ministry of Education, 1984): "In the past, the foremost aim was to teach a trade in order to train the pupil to execute the work and to use materials and techniques employed daily by those in their immediate environment... today, technology education in elementary schools focuses on providing the pupil an understanding of the design and production of a product." In a successive reform made in the mid–1990s, technology studies in elementary and junior high schools were integrated into the comprehensive science and technology program mentioned above.

The NTAIE report (2004) proposed restructuring the educational system back again to only two sectors: elementary school (Grades 1–6) and secondary school (Grades 7–12). As a part of this new structure, it is expected that technology studies in elementary schools will remain part of the integrated science and technology program. Another probable change is that technology studies in secondary schools, as an independent discipline, will be targeted primarily at high-achieving pupils and focus on topics closely related to engineering, science and mathematics.

The conceptual changes that technology study has undergone over the past three decades can be exemplified by the way the concepts of 'control' were taught within the K–12 education system, as follows:

- Since the second half of the 1960's, principles of control were a part of technology studies only at the secondary level. Until the end of the 1970's, theoretical studies were less extensive than practical studies. Students were engaged primarily in constructing systems based on mechanical controllers and sensors, instrumentation control of temperature or liquid flow in the chemical industry. This was a semi-vocational framework where some teachers had an industrial background.

- The rapid development of technology, in particular electronics fostered at the end of the 1970's, involved the introduction of the theory of control systems as a part of technology studies. For the first time, the curriculum included the learning of theoretical concepts such as feedback, steady-state response, transient response and stability of control systems. In the laboratory, students set-up and experienced control systems such as speed and position of electro-mechanical mechanisms, using electronic sensors, amplifiers and motors.

- During the 1980's, digital systems became central to technology, and gained an increasing role in technology studies. Students used discrete digital components and microprocessors for the design of digital control systems such as traffic lights or mini-robots. Since personal computers became available in the technology laboratories, a growing portion of students' lab work and projects have included hardware and software. The study of theoretical concepts, like sampling, analog-digital conversion and algorithms for continuous closed-loop control, became an integral part of the secondary school technology curriculum.

- Since the early 1990s, the design and construction of sophisticated autonomous robots has captured the attention of talented pupils from a range of schools, including those which traditionally did not offer classes for technology majors. Teams of pupils have competed in national and international robotics contests with good achievements.

The concept of control was a part of technology education in junior high school since the beginning of 1970's. The study of this concept has developed gradually during the years, but remained mainly an observable phenomenon which was achieved through experiences rather than a concept theoretically-based on mathematics definitions. The study of concepts such as feedback control and sensing was also introduced into the integrated program of science and technology for junior high schools from 1996. The shift from teaching instrumentation for secondary high school students, to teaching concepts of control systems within the compulsory program for science and technology at junior high school and basically also to the primary school pupils, reflects the change that technology studies has undergone in Israel. Presently, concepts like control or design are included in the curriculum aimed at all K–12 students.

OVERVIEW OF TECHNOLOGY TEACHER EDUCATION

Technology teacher training in Israel is carried out within three frameworks:
- Pre-service training, undertaken within regional academic colleges or in universities.
- In-service training programs, taken in regional and national teachers' centers, operated mainly by universities.
- Graduate studies (M.Sc. and Ph.D.) for teachers, completed in the universities.

As Israel is a relatively small country, the number of new technology teachers required each year is not substantial. However, the in-service programs for teachers and tracks for graduate studies have become a major ingredient of the teacher training system for several reasons:
- Frequent reforms in the curriculum.
- Pressure on schools to raise teaching standards.

- Need to introduce updated teaching-learning methodologies into schools.

Consequently, this chapter pays attention not only to technology teachers' pre-service training, but also to frameworks for in-service training and advanced studies for teachers.

Elementary and junior high school teachers are trained primarily in regional academic colleges, some of which are special colleges for education and others are colleges that offer teacher training programs. Senior high school teachers are trained primarily in the country's major universities, which have teacher training programs in their education departments. There is a marked difference between the teacher training programs in each of these frameworks. The college program emphasizes the educational-pedagogical aspects of teacher training and places great weight on courses in the history and philosophy of education, teaching methods, or introducing information and communication technologies into the school. The training involves a significant amount of practice by the student teacher under the supervision of the college staff. Technology teaching students in universities study towards a full bachelor's degree in engineering, which also provides them with a qualification for an Academic Teaching Certificate. The university program emphasizes disciplinary studies and places less weight on courses in education and teaching.

Drawing on Shulman's (1986) model, it may be said that teacher training in colleges places greater emphasis on the pedagogical knowledge required for the teachers and less on content knowledge; teacher training in universities concentrates more on content knowledge and less on pedagogical knowledge. Teacher training in colleges is often criticized since teachers who are graduates of these institutes are sometimes blamed that they lack sufficient knowledge in the content areas they are supposed to teach. Therefore, claims are being made that these teachers do not contribute enough in raising pupils' scholastic achievements, fostering excellence and strengthening the teacher's status. Teacher training in universities is frequently criticized since, even though the graduates might be good knowledge disseminators, their educational expertise is limited. This might, for example, cause difficulties with teaching in heterogeneous classes or in responding to the needs of diverse populations.

The teacher training system has been exposed to conflicting pressures placed on education over recent years. In light of the heightened socio-cultural conflicts and growing alienation and violence, there has recently been a growing voice/need to strengthen the teacher's pedagogical role and provide more critical thinking in the teacher training program (Giroux, 1989). These ideas emphasize one of the teacher's main roles, which is to educate pupils about democratic values and equality, and to impart values opposing exploitation or inequality related to status, gender, ethnicity, or communication. On the other hand, increasing pressure is being placed on the education system to raise pupils' scholastic achievements in accordance with a well-defined set of standards and strict discipline within the schools.

The current reform proposed for the Israeli education system (NTAIE, 2004) emphasizes the importance of teachers' knowledge of subject matter and adopts the university program as the preferred teacher training framework. Thus, one can expect that in the future most technology education teachers will be graduates of an engineering faculty. However, teacher educators will have to pay much closer attention to developing what Banks and Barlex (1999) call 'school knowledge,' which serves as an intermediary between subject knowledge and pedagogical knowledge. The challenge is to educate teachers to comprehend that the aim of technology education is not to simply teach content in electronics or mechanical engineering, but to impart to pupils "research and learning abilities, cognitive and personality capabilities that will make it possible for the child to be learning during his entire life" (NTAIE, 2004).

STRUCTURE OF TECHNOLOGY TEACHER EDUCATION

Pre-Service Training

Teaching students in Israel study a four-year program in colleges of education towards a Bachelor of Education (B.Ed). degree in Science and Technology teaching. Teachers with diverse backgrounds study within this framework to teach subjects such as 'design and products' and 'technological systems and control' using the Science-Technology-Society (STS) approach. In the university program, which will probably become the model for teacher training, the student-teachers study towards a full bachelor's degree in their discipline of specialization. In mathematics or science, this is a Bachelor of Arts (B.A) degree, which consists of about 120 learning units (three years); in technology this is usually a Bachelor of Science (B.Sc) degree in one of the engineering fields, which consists of about 160 learning units (four years). In addition, student-teachers are required to study a program for an Academic Teaching Certificate, which consists of 30-35 learning units (about one academic year). This program consists of courses such as:

- History and Philosophy of Education
- Educational Psychology
- Social Psychology
- Teaching Skills
- Methods for Teaching Technology
- Teaching Advanced Subjects in Engineering
- Teaching the Integrated Program for Science-Technology-Society (STS)
- School Practicum

Both technology and science student-teachers study one or two courses together that are related to teaching the STS-oriented program aimed at young children, as outlined later in this chapter. Over 90% of teachers in university pre-service teacher training programs devote most of their time to subject matter or general courses in education, rather than courses on teaching technology. In addition, the school practicum in the university pre-service teacher training program is short, for example eight lessons guided by an experienced teacher together with the university lecturer. This explains the importance of ongoing efforts in fostering the professional development of teachers on the job.

In-Service Training

Over the past decade, the Israeli educational system has witnessed considerable investments in in-service training programs for science and technology teachers. These programs were aimed at preparing teachers on duty to teach the integrated science and technology curriculum and to update their pedagogical knowledge in subjects such as project-based learning and the use of computer technologies in the class. In many cases, the in-service training programs preceded the introduction of the new curriculum into pre-service programs in the universities or was run concurrently, for several reasons:

- Developers of new curricula regularly include experienced teachers in the teams, who prepare the new learning materials and run pilot studies in schools. In Israel, this is a standard requirement for academic bodies engaged in developing new curricula sponsored by the State.
- Educational experts acknowledge that skilled teachers are the key people behind any educational improvement. For instance, outstanding teachers often help new teachers during their first steps in teaching.
- Veteran teachers frequently hold central positions in their schools, such as heads of science and technology departments, or laboratory coordinators. The commitment of these teachers to introducing new curricula or ongoing improvements is crucial.

Graduate Studies in Science and Technology Education

Recently, there has been a growing demand for post-graduate studies among science and technology teachers. Almost all major Israeli universities offer programs for masters and doctorate degrees tailored specifically to teachers in the fields of mathematics, science, technology and computers. All universities require that graduate study candidates must have a B.Sc, B.A or B.Ed degree in their field of specialization, an Academic Teaching Certificate, and at least two years of teaching experience following the internship stage. As a typical example, the graduate program for science and technology teachers at Ben-Gurion University of the Negev (BGU) follows:

The declared objectives of the BGU graduate program are: (1) "To provide the teachers with a theoretical and practical framework for their professional development," and (2) "To give the teacher the opportunity to develop into a teacher-investigator and professional, researching both himself and his pupils, and enhancing his educational work with the knowledge gained from his research studies"(BGU, 2004).

The program offers two study tracks:

- A research track, which consists of studying courses for 33 credit points and preparing a final thesis.
- A general track, which consists of studying courses for 39 credit points and preparing a literature survey which is a mini-research project, on a selected subject in science and technology education.

The curriculum comprises three main areas: educational research, advanced subjects in science and technology, and the history and philosophy of science and technology. The students study courses such as:

- Research Methods and Paradigms in Science and Technology Education
- Alternative Methods in Educational Evaluating
- Teaching and Learning in a Computerized Environment
- History and Philosophy of Science and Technology
- Advanced courses in specialization areas, such as mathematics, biology or engineering.

Teachers who take three courses each semester can complete their studies in about five semesters.

Graduate studies for teachers are becoming an increasingly important framework for fostering teachers' professional development. Since most of the teachers who opt for advanced studies are beyond the 'battle for survival' of the first few teaching years, these teachers expect a broader outlook on education and educational research and are less interested in learning how to teach a specific curriculum.

AN EXAMPLE: ISRAEL INSTITUTE OF TECHNOLOGY

Undoubtedly, cooperation between science and technology teachers in implementing the combined science and technology curriculum has been one of the central issues concerning scholars and decision-makers. This section presents two examples of efforts to promote such cooperation.

Frank (2004) presents an example of integrating Project-Based Learning (PBL) as the central axis of a pre-service teacher training course at the Technion, Israel Institute of Technology. Participants were 46 teaching students who were studying towards an Academic Teaching Certificate for technology or science, in parallel to their studies towards a bachelor's degree in one of the science or engineering faculties. The course was presented by an interdisciplinary teaching team comprising an expert in biology teaching, an expert in technology teaching and a teaching assistant holding a degree in chemistry/biology teaching. The participants were taught two parallel models of project-based learning (Frank and Barzilai, 2004):

- **Project-based science** according to the inquiry approach, involving "presenting a research question, formulating scientific predictions, designing and conducting an investigation, gathering and analyzing

data, making interpretations and identifying alternative explanations, drawing conclusions."
- **Project-based technology**, which adopts the design approach, involving "identifying a need, determining system requirements, collecting and analyzing data, conducting a feasibility study, examining alternative solutions, choosing an optimal solution, designing the system, producing and testing a prototype/physical model, presenting outcomes."

Course participants worked in compound science/technology pairs on projects of their choice. The evaluation study showed that this type of pre-service teacher training exposes future teachers to the advantages and limitations of project-based learning in science and technology, and fosters cooperation between teachers in both subjects. According to Frank and Barzilai (2004): "While performing the projects the students were exposed to the benefits of PBL: Gaining interdisciplinary knowledge, becoming familiar with the nature of engineering and design, becoming aware that engineering design operates within constraints, realizing that in engineering there is always more than one possible solution, understanding that science and technology are strongly connected, experiencing the importance of the cooperation between the team members, acquiring knowledge through active and experiential learning, taking responsibility for the learning, acquiring communication skills and methods of decision-making within a team, and enhancing of one's self-esteem."

Introducing project-based learning in teachers' pre-service training programs, as demonstrated by the above example, is important for exposing the trainees to the issue of linking the teaching of science and technology. However, promoting teamwork among experienced teachers through in-service training programs, as shown in several examples below, is no less important, since veteran teachers are very familiar with the school context on the one hand, but tend to concentrate on their field of specialization, on the other.

Barak and Pearlman-Avnion (1999) report a case of introducing the The Sound of the System learning unit in junior high schools. On the scientific side, pupils studied subjects such as electricity, sound and waves, including theory and experimentation. On the technological side, they learned concepts such as system, amplification and design, and built an electronic audio amplifier. Barak and Raz (2000) presented a program in which talented junior high school pupils studied physics and technology during one academic year around the design and construction of electronically-controlled hot-air balloons. The same teachers ran an Amusement Park program in their schools, whereby junior high school pupils worked in teams on the design and construction of small electro-mechanical models of systems, such as a computer-controlled carousel or a Ferris wheel. In these three examples, teacher training consisted of regional in-service courses as well as mentoring the teachers at school. Science and technology teachers studied together for about four hours a week throughout the school year, and fulfilled all the scientific and technological tasks set for their pupils.

Some conclusions drawn from these programs were:

- Joint science-technology projects, especially the in-service training courses, broke the ice between the science and technology teachers, who had rarely worked together before.
- Science teachers admitted that they visited a technology laboratory in their school for the first time, and were able to get an idea of what pupils learned in technology. Technology teachers expressed a similar opinion.
- Although teaching a joint program, the science teachers adhered closely to the disciplinary field in which they specialized, such as physics, chemistry or biology.
- The technology teachers adhered closely to the disciplinary field in which they specialized, such as electronics, computers, mechanics or robotics.

A broad scientific-technological educational project might place heavy demands on qualified teachers. It is unrealistic to expect a teacher to possess sufficient expertise in both domains to ensure that the entire project could be completed by one teacher.

TEACHER CERTIFICATION

As previously mentioned, there are three tracks for training technology teachers in Israel, each of which provides a different certificate. The first track is studying towards a Bachelor of Science (B.Sc) degree in Engineering, plus a specific program consisting of courses in education and teaching. This track is also suitable for engineers in industry wanting to integrate into the educational system at a later stage in their careers. The second track is studying towards a Bachelor of Science (B.Sc) degree in Technology Education. Teachers in this track also specialize in one of the engineering fields. The third track is studying towards a Bachelor of Education (B.Ed) degree, which is offered by academic colleges of education. All of the above tracks grant teachers an Academic Teaching Certificate, which automatically provides them with a teaching license.

Several suggestions were recently brought up for discussion aimed at raising teachers' educational and professional status. One of the main recommendations of the NTAIE committee (2004) was that all teachers would have to complete a full B.Sc or B.A degree in the discipline they would teach. It is hoped that if the B.Ed degree is no longer offered, this will not only raise teachers' professional knowledge levels, but also facilitate a greater mobility among teachers from education to industry, or vice versa. Another suggestion is that new teachers complete a one-year internship and pass a certification exam. In this way, the teaching certification process would be similar to that in other areas such as medicine or law. Furthermore, it is proposed to enforce certification exams, or introduce another evaluation model, for teachers every few years and/or prior to their receiving a promotion or a tenured position.

Issues regarding teacher training, professional development and certification are not unique to Israel, and other countries deal with them as well (Oliver and McKibbin, 1985: McCaslin and Parks, 2002). Decision-makers are aware that raising the certification requirements of teachers could worsen the problem of the lack of qualified teachers in the short-term, but it is hoped that these changes would strengthen teachers' status and attract outstanding teachers in the long-term.

CONCLUSION

During the last two decades of the 20th century, many teachers, teacher educators, curriculum developers and researchers took part in an extensive and comprehensive debate on all aspects concerning teaching and learning technology in the Israeli education system. Issues regarding the philosophy of technology education, or questions such as 'What is technology?' and

'What are the objectives of teaching technology to young children?' stood at the middle of this debate. In this context, the well-known international gathering that took place in Israel, JISTEC '96, was also a part of the struggle to clarify the objectives and methods of technology education in schools (Tamir and de Vries, 1997).

Since the beginning of the new millennium, the Israeli education system has been undergoing extensive and significant reform, which will increasingly affect all current frameworks of teachers' pre-service training, certification and professional development. Although the broad consequences of this reform on training technology teachers are not clear at the time of writing this chapter, it is quite obvious that one of the central recommendations is that each teacher will have to acquire at least a bachelor's degree in a specific area of science or engineering. Since no academic degree is offered in technology, it is apparent that the majority of future technology teachers will need an engineering background. The question is to what extent, and how, will the system prepare teachers who value the aim of technology education to convey to pupils the knowledge and skills essential for integrating them successfully into the future society and to contribute to their intellectual development, rather than teaching specific skills in electronics, computing or mechanical engineering. The important role of technology education in fostering pupils' higher intellectual skills has been observed in several follow-ups on technology education outcomes in Israeli schools. The Israeli experience has shown that competent pupils who choose to study technology expect to cope with sophisticated tasks. To address this challenge, pre-service and in-service training programs for science and technology teachers will have to deal with a range of issues, such as developing project-based learning in technology, fostering teamwork among teachers having diverse backgrounds and expertise, and strengthening the links between technology studies and sophisticated industry, academia and the community.

REFERENCES

Advisory Committee on Pupil Financial Assistance (ACSFA) (2000). Factors affecting access in the twenty-first century: a round table discussion of early intervention, remediation, and support services, Burlington, Vermont: University of Vermont. http://www.ed.gov/about/bdscomm/list/acsfa/sept00brf.doc

Banks, F. & Barlex, D. (1999). No one forgets a good teacher! – What do 'good' technology teachers know? *The Journal of Design and Technology Education*, 4 (3), 223–229.

Barak, M., Yehiav, R., Mendelson, N. (1994). Advancement of low achievers within technology studies at high school, *Research in Science and Technological Education*, 12(2), 175–186.

Barak, M. & Raz, E. (2000). Hot-Air Balloons: project-centered study as a bridge between science and technology education, *Science Education*, 84 (1), 27–42.

Barak, M. (2004). Issues involved in attempting to develop independent learning in pupils working on technological projects, *Research in Science and Technological Education*, 22(2), 171–183.

Barak, M. & Pearlman-Avnion, S. (1999). Who will teach an integrated program for science and Technology in Israeli junior high schools? A case study, *Journal of Research in Science Teaching*, 36 (2), 239–253.

Barak, M. & Tamir, A. (2003). Technology education in Israel – Aiming to develop intellectual abilities and skills via technology studies. In: G. Graube, M. J. Dyrenfurth, and W. E. Theuerkauf (Eds.), *Technology Education, International Concepts and Perspectives*. New York: Peter Lang Press, 221–228.

Ben-Gurion University of the Negev (BGU) (2004), Graduate Program for Science and Technology Education. http://www.bgu.ac.il/SciTecEdu/

Coles, A. S. (1999). School to college transition programs for low income and minority Youth, *Advances in Education Research*, 4, Washington, D.C.: National Library of Education, Office of Educational Research and Improvement, U.S. Department of Education.

Doppelt, Y. & Armon U. (1999). LEGO-Logo (Multi-Techno-Logo) as an authentic environment for improving the learning skills of low-achievers, Paper presented to the *EUROLOGO Conference*, Sofia, Bulgaria, August.

Drucker, P. F. (1994). The age of social transformation, *The Atlantic Monthly*, November, 55–80.

Forsyth, A. J. & Furlong A. (2003). Access to higher education and disadvantaged young people, *British Education Research Journal*, 29 (2), pp. 205–225.

Frank, M. and Barzilai, A. (2004). Project-Based Technology in the Science and Technology Curriculum: A Teaching Approach for Developing Technological Literacy, *Paper presented at the National Association for Research in Science Teaching (NARST) Conference*, Vancouver BC, April, 1–4.

Fullan, M. (1994). Coordinating top-down and bottom-up strategies for educational reform, *Systemic Reform - Perspectives on Personalizing Education*, September http://www.ed.gov/pubs/EdReformStudies/SysReforms.

Giroux, H. (1989). *Schooling for Democracy, Critical Pedagogy in the Modern Age*. London: Routledge.

McCaslin, N. L. & Parks, D. (2002). *Teacher Education in Career and Technical Education: Background and Policy Implications for the New Millennium*, The National Career and Technical Teacher Education Institute, The Ohio State University. http://www.nccte.com.

Ministry of Education and Culture (1984). *The Technological Education in Israel*, Jerusalem.

Ministry of Education and Culture (1996). *Science and Technology Studies for Junior High School*, Jerusalem: Author.

National Task-Force for the Advancement of Israel's Education (NTAIE) (2004). Jerusalem: Ministry of Education and Culture.

Oliver, B. & McKibbin, M. (1985). Teacher Trainees: Alternative Credentialing, *Journal of Teacher Education*, 36 (3), 20–23.

Shulman, L. S. (1986). Those who understand: Knowledge growth in teaching, *Educational Researcher*, 15(2), 4–14.

Tamir, A. & De Vries, J. M. (1997). Preface, *The International Journal of Technology and Design Education*, 7 (1–2), 1–2.

Third International Mathematics and Science Study (TIMSS) (1999). International Study Center, Boston College, Lynch School of Education. http://timss.bc.edu

Yager, R. E. (Ed.) (1996). *Science/Technology/Society as Reform in Science Education*, Albany, NY: State University of New York Press.

Technology Teacher Education in Japan

Chapter 7

Hidetoshi Miyakawa
Aichi University of Education, Japan

INTRODUCTION

Japan consists of 47 prefectures that include 2,395 autonomous municipalities (2005). Local residents elect members of their local council and the head of their local administration (prefectural governor; or mayors, town managers, or village headmen) and committee members of the board of education are appointed. Municipal corporations are endowed with a broad jurisdiction in terms of the local administration, whereas the main functions of branch offices of the national administration are to provide the municipal corporations with advice and assistance. However, national ministries exert influence on local administrations through the establishment of ordinances and standards.

In terms of the national funding system for school education, the nation provides financial support for about half of the expenses of compulsory education including teachers' salary, construction of school buildings, teaching materials/teaching tools, economic aid for attending schools, and educational promotion in remote places. In short, the budget is covered half by the nation and half by the prefecture or the municipality, and so the involvement of the nation is significant.

Historically, technology education was provided through drawing and handicrafts in elementary schools, but now, at lower secondary schools, Industrial Arts and Homemaking Education has been taught as a required subject since 2002 for 70 hours in the first school year, 70 hours in the second year and 35 hours in the third year. Previously, the subject names Industrial Arts, and Homemaking Education, were handled within a single framework under the policy that males would take industrial arts and females would take homemaking education. Today, however, all students take the same curriculum without distinction of gender. The number of hours for industrial arts education and homemaking education is 35 in the first year, 35 in the second year and 17.5 for the third year, representing a significant reduction in hours when compared with the previous curriculum.

With regards to the educational content of technology education in the new curriculum, the six areas that had comprised the previous curriculum: Woodworking, Metalworking, Electricity, Machines, Cultivation, and Fundamental Education of Information Processing were integrated into two content areas in the new curriculum, viz., Technology and Manufacturing, and Information and Computer. The Technology and Manufacturing curriculum consists of the following content:

1. Roles of technologies
2. Design of manufactured articles
3. Material processing technology
4. Mechanism and repair of devices
5. Energy conversion
6. Cultivation of crops

The Information and Computer curriculum consists of the following content:

1. Roles of information means
2. Bases of computer
3. Use of computer
4. Telecommunication network
5. Use of multimedia
6. Program and instrumentation/control

Fundamental Education of Information Processing, which was one of six areas in the previous curriculum, was largely dealt with in the new curriculum area of Information and Computer, with the remaining five areas integrated into Technology and Manufacturing. This structure of the curriculum reflected the advent of the information society. However, when considering that Japan achieved the present economic growth through manufacturing after World War II and that Japan will continue to be a nation active in manufacturing, this rationale for curriculum composition will remain an issue. Nevertheless, it is encouraging that the number of hours of the teaching program is left to the judgment of the school or the teacher, and there are reportedly many schools in which much time is allocated to the subject Technology and Manufacturing.

In these two subjects, the content areas from 1 to 4 are compulsory, and 5 and 6 are elective. The teaching content and method are, however, not uniform because these are designed by the community or school. On the other hand, in the new curriculum introduced in 2002, periods of integrated study have been set, and are taught for 70–100 hours in the first school year, 70–105 hours in the second school year and 70–130 hours in the third school year. The main educational content for these classes includes; Information, Environment, International Understanding, and Welfare, and emphasis is placed on hands-on learning as seen in manufacturing. Therefore, it is quite likely that technology education is promoted in this way, being extremely significant in light of both the educational content and method.

In terms of school education, samples of textbooks used by the students are prepared by textbook companies based on the Courses of Study notified by the Ministry of Education, Culture, Sports, Science and Technology and textbooks that pass their assessment are permitted to be sold. Although the adoption of textbooks is left to the discretion of boards of education of prefectures or municipalities, expenses for the textbooks are all borne by the nation. Therefore, all textbooks used in all schools all over Japan have been examined by the Ministry of Education, Culture, Sports, Science and Technology, rendering a more uniform education all over Japan. There are two textbook companies for the industrial arts and domestic science in Japan, and these provide the specified texts.

Technology education in upper secondary schools is divided into general education (general course) and professional education (industrial course). The general course is for students who wish to proceed to higher education or who have not decided on specific vocational areas after graduation, and it includes the information subject which started in 2003, that is related to technology education. However, a technology teacher cannot take direct charge of the information course without obtaining an additional certificate.

The industrial course is subdivided into many areas including technical, agricultural, business, and fishery. The certificate for industrial education, obtained by many technology teachers, enables them to be involved in all industry-related education including the machine course, the electricity course and the architecture course. Approximately 25% of upper secondary school students across Japan are enrolled in the industrial course.

In addition, the comprehensive course in upper secondary school, which provides both general and professional education, was established for upper-secondary school in 1994. Although subjects in the comprehensive course can be selected in accordance with diversified interests, ability, aptitude, and desired future interests, and from a wide variety of subjects, the present number of schools that offer the course is still small, but is expected to increase.

HISTORY

The source of Industrial Arts and Homemaking Education goes back to the Fundamental Law of Education and the School Education Law that were established in 1947. The new education system's lower secondary school was established under the 6–3–3–4 school system, and the 'vocational course' started as a subject corresponding to the "cultivation of the ability to select the future course in accordance with basic knowledge and skills in occupations needed in the society, an attitude respecting labor and the personality," provided in Article 36 (2) of the School Education Law. This course consisted of five subjects – agriculture, industry, business, fishery, and domestic science, and students were expected to take one or several subjects to study for 140 hours. Female students took homemaking education.

From 1951 onward, the formerly separated Homemaking was integrated into a single subject with the name altered to Vocational & Homemaking Education. The Course of Study consisted of four classes and 12 items: the first class (cultivation, breeding, fishery, food processing), the second class (handicrafts, machine operation, drafting), the third class (paperwork, business bookkeeping, calculation), and the fourth class (cooking, sanitation and child-care), and it was mandated that students study these for 140 hours in every school year. In 1951, the Central Vocational Education Council was established to study the nature, goals and content of Vocational& Homemaking Education.

From 1957 onward, the content was divided into six groups: agriculture, industry, business, fishery, homemaking education and vocational guidance. It was mandated that every group, except fishery, was compulsory for at least 35 hours for all students, and the fishery subject was elective. Around 1957 was an era of international technological innovation, symbolized by the former USSR's artificial satellite Sputnik, when rapid scientific and technological advances brought about economic stability and abundance to Japan, and the public calls for the promotion of scientific and technological education intensified.

From 1962 onward, the Vocational& Homemaking Education Course was reformed to establish the male-oriented industrial arts and the female-oriented homemaking. The male-oriented education consisted of Designing/Drafting, Woodworking, Metalworking, Cultivation, Machines, and Electricity. The female-oriented education consisted of Cooking, Clothing Manufacturing, Household Handicrafts, Household Machines, Household Electricity, and Child Care with the individual details specified by the grade. The number of class hours was three per week, or 105 per year totaling 315 hours in the three years of lower secondary school.

Later, in order to cope with the scientific and technological advances and rapid economic, social and cultural progress, it was determined that, from 1972, the nature of the practical subjects Industrial Arts and Homemaking Education be clearly defined. The goal was that "the male-oriented education and the female-oriented education be associated with each other and basic items be carefully selected with the range and the degree clearly defined" in terms of the educational content, and "consideration be given so that guidance may be provided flexibly in accordance with the actual conditions of the community, the school, and the students". For the male-oriented education, the comprehensive practice was abolished, cultivation was transferred from first year to the third year, and design was integrated into all subjects with drafting left as a separate area. As for the female-oriented education, household handicrafts were abolished, dwelling was newly established, clothing manufacturing was changed to clothing, and cooking was changed to food. The number of class hours was the same as the previous Course of Study.

Although the goal of Industrial Arts and Homemaking Education from 1981 basically remained unchanged, the educational content that supported the goal changed significantly. The conventional male-oriented and female-oriented educational groups were abolished and the new educational groups were composed of 17 areas: Woodworking 1 and 2; Metalworking 1 and 2; Machines 1 and 2; Electricity 1 and 2; and Cultivation for males, and Clothing 1, 2 and 3; Food 1, 2 and 3; Dwelling; and Child Care for females. It was decided that a school should select a total/maximum of seven areas to offer to its students on the so-called mutual entry basis, based on the school's judgment: six areas or more from the industrial arts group and one area or more from the household science group to male students, and five areas or more from the household science group and one area or more from the industrial arts group to female students. The number of class hours became 70 hours each year for the first two years and 105 hours for the third year, totaling 245 hours in three years.

The midterm year for the decade of International Women's Year was 1980, when equal employment opportunity and equal curriculum opportunity were raised as issues in the promotion of the International Convention on the Elimination of All Forms of Discrimination against Women. Partly because of these circumstances, it was decided in 1993 that Woodworking, Electricity, Cultivation, Home Life, and Food would become required subjects and taught under the same curriculum for males and females, drawn from the six industrial arts group areas consisting of Woodworking, Metalworking, Machines, Electricity, Cultivation, and the newly established Information Processing, and the five areas composed of Clothing, Food, Dwelling, Child Care and newly established Home Life. In addition, it was mandated that students study seven of the eleven areas, with the number of class hours being 70 in the first year, 70 in the second year and 70–105 in the third year. Woodworking and Home Life became standard subjects for study in the first school year. Industrial Arts and Homemaking Education were added as one of selected subjects, with a maximum of 35 hours permitted to be allocated in the third school year. In the meantime, in order to respond to the newly established Fundamental Education of Information Processing, teacher training by the Ministry of Education, prefectures/municipalities, research institutes or private companies was actively implemented from about 1985.

Trends in advancing technologies; computerization and internationalization were reflected in Industrial Arts and Homemaking Education from 2002 onward. Woodworking and Electricity, which were required areas in the previous Course of Study, and Metalworking, Machines, and Cultivation, which were selective areas, were integrated into Technology and Manufacturing. Information Technology, which was a selective area, became Information and Computer. In the homemaking education field, food and home life, clothing and dwelling were integrated into the homemaking education field as Independent Living and Clothing, Food and Housing, and Home Life and Childcare were integrated into Family and Home Life.

OVERVIEW OF TECHNOLOGY TEACHER EDUCATION

Technology teacher education did not parallel the curriculum changes. Technology teachers are trained mainly in the engineering or agricultural department of national or private universities or in a teacher's college. The former provides credits in the subjects required, based on the Educational Personnel Certification Law. However, in the teachers' colleges, credits cannot be provided by specialized instructors in all subjects of Woodworking, Metalworking, Electricity, Machines, Cultivation, and Information due to low numbers of instructors resulting from a decrease in the number of students. The teachers are complemented by support from part-time or non specialized instructors. Students who aim to be technology teachers need to acquire skills in a broad range of domains and so they have no choice but to study broadly and shallowly.

On the other hand, those students who obtain a teachers certificate in such faculties as electrical or mechanical engineering, or timber engineering in the agricultural department, study deeply in the specialized fields but relatively shallowly in the other fields.

In general, there are two paths to becoming a technology teacher: one is to enter a university with the objective of becoming a technology teacher from the beginning, and the other is to obtain the technology teacher certificate as a second certificate in addition to the first objective of becoming a teacher for elementary school or another subject. However, the latter students mostly graduate from their universities with a small number of credits in specialized content areas associated with the industrial arts, or with a second class certificate, and as a result, lack expertise and skills.

Increasing numbers of female students have recently entered universities with the objective of obtaining the teacher certificate for technology education, whereas recruitment tests tend to channel them into elementary school teaching. It is easier to get a job as a teacher at elementary school compared to lower secondary school. The number of female technology teachers in Japan is less than one percent.

The continuing professional development of in-service teachers has recently become an important issue in Japan. Local boards of education provide training programs for beginning teachers in their first year of teaching, and teachers who have ten years of experience. However, as teachers are granted permanent employment and professional development is not compulsory, it is left to the discretion of teachers. Active teachers belong to a national or international research institute or professional organization and attend workshops or conventions for professional discussions and to make presentations. Incidentally, there is a nationwide organization for teachers of Industrial Arts and Homemaking Education for lower secondary schools in Japan, which holds workshops in different prefectures every year. Research institutes with university instructors as the nucleus include the Japanese Society of Technology Education, which implements research activities through its publications and holds general assemblies and workshops, as well as being closely aligned with technology education at lower secondary schools.

STRUCTURE OF TECHNOLOGY TEACHER EDUCATION

The requirement for technology teachers in Japan is a bachelor's degree from a higher education institution on completion of four years study, after graduation from an upper secondary school. There are eleven single-department teachers' colleges and 36 teachers' departments of universities in Japan that train technology teachers. As of 2003, there are 281 instructors in technology education courses. In terms of their educational background, there are 196 experts in technology, 47 instructors in charge of educational content and 38 experts in technology education. The number of teachers who graduated from the universities in 2003 was 329, comprising 250 males and 79 females. The number of technology education students in post-graduate master's courses is 85 male and eleven female students. Since the education system in Japan is uniformly regulated, the educational content provided in all the teacher's colleges/departments is very similar. Aichi University of Education will be used later in this chapter to exemplify the programs.

The subjects in the bachelor's degree for training technology education teachers for lower secondary schools are as follows (number of credits in parentheses):

- Common subjects (29)*

  Liberal arts   Constitution of Japan ......(2)
                                    Basic subject ....(6)
                                    Theme subject ....(8)

  Introduction of information education ....(2)

  Foreign language subjects   First foreign language ......(4)
                                         Second foreign language ....(2)
                                         English communication ....(2)

  Sport subjects (3)

- Professional education subjects (83)

  Introductory Subject for Professional Education ....(2)
  Common major subject ....(2)
  Course study subject ....(2)
  Course major subject ....(34)
  Teaching profession subjects
                               Course educational subject ...(8)
                               Educational subject ....(25)
  Subject related to the course or teaching profession ....(4)
  Graduation study ....(6)

- Independent subject ....(16)

  Total number of credits   (128)

*The hours depend on the subject.

The dedicated/specific subjects for technology teachers are course major subjects to learn the specialized fields, course education subjects to learn educational methodologies and educational subjects. These are described in detail as follows:

(1) The course major subjects provide breadth in specialized fields related to industrial arts, and targets the following six required areas to obtain the teacher certificate. The course offers 18 required subjects with 20 credits and 24 elective subjects with 31 credits, totaling 51 credits.

- Woodworking (drafting, lectures and practice on woodworking)
- Metalworking (lectures and practice on metalworking)
- Machines (lectures and practice on machines)
- Electricity (lectures and practice on electricity)
- Cultivation (lectures and practice on cultivation)
- Information Technology (lectures and practice on computers)

There are three types of license for technology teachers; the first, second and special grade. The first grade license, which is the level of bachelor's degree, requires 40 credits of the subjects in the major course. The second license, which is the level of an associate's degree, requires 20 credits. The special license requires 6 credits taken at a graduate school. However, required credits for becoming technology teachers changed for those students who entered universities from 2000, as shown in Table 1.

Required credits of the subjects in the major course for the first grade license are going to be reduced from 40 to 20 credit hours, which is more than one credit hour for each subject. Similarly, ten credit hours will be required for getting the second grade. This is because more credits will be allocated to education subjects.

Table 1.
Credits Required for the License of Technology Teacher.

Program	Subject Area	Minimum Credits (1990–1999)	Minimum Credits (2000–)
Industrial Arts	Woodworking[*1]	6 or 4	1
	Metalworking[*1]	4 or 2	1
	Machines[*2]	6 or 4	1
	Electricity[*2]	6 or 4	1
	Cultivation[*2]	2	1
	Information Technology[*2]	2	1
		Total 40[*3] (20)[*4]	Total 20[*3] (10)[*4]

*1: including drawing and practical exercises   *2: including practical exercises.
*3: The first grade license.   *4: The second grade license.

(2) Course education subjects cover the methodologies for technology education. They include the following four required/compulsory subjects of two credits each.

- Technology Education C I
- Technology Education C II
- Technology Education C III
- Technology Education C IV

Technology Education C I covers topics in relation to various environments surrounding technology education and studies them from a broad perspective. The course covers the educational administration in technology education; the history of technology education; the value of technology education; the environment and technology education; computerization; internationalization; psychology of learning and teaching materials.

Technology Education C II covers instructional plans; learning processes; various tests; preparation of survey forms; selection and development of teaching materials; development of learning processes; usage of teaching materials; analysis of classes; and cultivates the sense of reality through the simulation of teaching practice.

Technology Education C III comprehensively studies purposes, contents, methods, and assessments of technology education while associating the past research results and the actual teaching practice. The course explains past research results relating to the goals, method, contents and assessment of technology education; various issues relating to selection, development, arrangement and application of teaching materials, and instructional design and educational guidance; and the study of theories and methods to solve those issues.

Technology Education C IV intensively studies important environmental conditions for implementing technology in the class and various assessments concerning the class, in a practical manner. The course deals with facilities and equipment, safety guidance, methods for educational assessment, assessment of contents/methods of learning and assessment of contents/methods of teaching.

(3) In the educational subjects, students will carry out student teaching in a lower secondary school. Aichi University of Education offers the following classes:
- Educational field study — basic student teaching
- Educational field study — major student teaching

- Educational field study — student teaching at a neighboring type of school
- Educational field study — applied student teaching

The main purpose of the basic student teaching is to deepen the understanding about basic activities of school education through observation and participation in classes and school events, to deepen the understanding about students and to cultivate positive attitudes toward the teaching profession through contact with children. The subject is optional and worth one credit.

The major student teaching is designed to understand the work and the mission of teachers, to recognize the overall structure of the school education system, to learn the methods and techniques for understanding and guiding children, and to further deepen the study in the students major. The subject is mandatory and worth five credits for four weeks of study.

The neighboring school experience is to learn the differences, characteristics, and continuity between school types through participation in a different type of school from the one where the students carried out their major student teaching. The subject is optional and worth two credits.

The applied student teaching is to cultivate professional competence and research, and development teacher capabilities through such activities as Team Teaching and Assistant Teaching, in order to further deepen the practical instruction. The subject is optional and worth one credit.

AN EXAMPLE: AICHI UNIVERSITY OF EDUCATION

Table 2 shows the list of subjects in the major course and the number of credits offered by the Department of Technology Education at Aichi University of Education.

Table 2.
Subjects at Department of Technology Education,
Aichi University of Education.

Subject Area	Title (Compulsory)	Credits	Title (Elective)	Credits
Woodworking	Woodworking method Woodworking practical exercise 1 Woodworking practical exercise 2 Drawing 1	1 1 1 1	Woodworking 1 Woodworking 2 Woodworking practical exercise 3 Woodworking experiment Drawing 2	1 2 1 1 1
Metalworking	Metalworking method 1 Metalworking practical exercise 1 Metalworking practical exercise 2	1 1 1	Metalworking method 2 Metalworking method 3 Metalworking practical exercise 3 Metalworking method 4 Metalworking experiment	1 1 1 1 1
Machines	Material Dynamics Mechanical Dynamics Mechanical experiment	1 1 1	Mechanical Engineering Mechanics Thermal Dynamics	2 1 2
Electricity	Electricity 1 Electricity 2 Electrical practical exercise	1 1 1	Electronics 1 Electronics 2 Electricity 3	2 2 1
Cultivation	Cultivation Cultivation practical exercise 1	2 1	Plant cultivation 1 Plant cultivation 2 Agriculture Cultivation practical exercise 2	1 1 2 1
Information Technology	Information 1 Programming 1 Basic Information Technology	2 1 1	Software 1 Software 2 Information 2 Programming 2	1 1 2 1
Technology	Industrial Arts 1 Industrial Arts 2 Industrial Arts 3 Industrial Arts 4	2 2 2 2		

For teaching practice, the university requires students to do teaching practice of a subject for four weeks in order to acquire the main teaching license for the subject. In order to acquire an additional teaching license for another subject, students need to spend a further two weeks teaching practice in that subject. Students are offered the opportunity to acquire various kinds of teaching licenses while at university. In addition to the course, a range of other activities are provided for the students and for practicing teachers. For teacher training in the future, it is important to provide opportunities during their study, for student teachers to know about and work with various technologies, and to communicate with incumbent teachers.

These courses are implemented after the Ministry of Education, Culture, Sports, Science and Technology approves an application made through the university. The broad theme consistently remains 'Course for Development of Teaching Materials for Technology Education' and a range of content is offered:

1. **Manufacturing of Wooden Products Using Thinned Wood.** The mountains of Japan are covered by natural and artificial forests. Some of the artificial forests are thinned and harvested periodically. As quantity of the thinned wood that can be actually used within society is limited, an attempt was made for thinned wood to be used in school education, associated with the issue of protecting the forest environment.

2. **Manufacturing of New Practice Work Materials Using Wood and Metal.** This is an attempt to manufacture new practice work materials by use of two types of materials – wood and metal. Manufacturing of creative products with previously unseen applications, designs and combinations was attempted.

3. **The Way What Teaching Materials Ought to Be Toward New Course of Study and Assessment.** The new Course of Study states that it is important to manufacture products not only by use of a single material but also by use of multiple materials. This course deals with manufacturing of products using three materials – wood, metal and plastic – that have not commonly been attempted so far.

4. **Robot Contest – Concept and Approach.** Robot contests have been held in various types of schools in recent years. This course proposes that, in order to implement a robot contest more effectively, it is important to push forward with manufacturing of robots and a contest after steadily mastering fundamentals/basics of materials.

5. **Toward New Course of Study Information and Computer.** The new Course of Study advocates that significant time should be allocated to Information and Computer technology. This course follows the curriculum from fundamentally required content to selective areas for development or application to enable smooth management of the class.
6. **Teaching Materials for Technology and Manufacturing linked with Integrated Study.** This course studies the methodology to link the industrial arts as required subjects, as elective subjects and as an Integrated Study effectively and efficiently through the manufacturing of practice work materials.
7. **Technology Education Advanced Theory I to Acquire Advanced Class Certificate.** Incumbent teachers who possess a first class regular certificate can acquire the advanced class certificate by accumulating a prescribed number of credits for lectures offered in graduate schools. This course was established to permit relevant teachers to acquire the credits through open lectures.

The above activities were offered by instructors, lecturers and students in collaboration with teachers and researchers outside of the university. The content has also received high evaluation from within the university.

TEACHER CERTIFICATION

Teachers from kindergarten to upper secondary level must possess a teacher certificate granted by a prefectural board of education in accordance with the Teacher Certificate Law established by the nation. Teacher certificates for lower and upper secondary schools are largely divided into three categories; a regular certificate, an advanced class certificate, and a temporary certificate for assistant teachers. Teachers who possesses a teacher certificate for kindergarten or elementary school teach all subjects while teachers with a teacher certificate for lower or upper secondary school teach the subject for which they were granted the certificate.

Teachers who wish to obtain certificates need to acquire credits in subjects relating to a prescribed course and the teaching profession, in a curriculum authorized by the Minister of Education, Culture, Sports, Science and Technology, in addition to a basic qualification. The regular teacher certificate is categorized into the advanced class certificate, the first class certificate and the second class certificate. Basic qualifications for teacher certificates are: acquisition of a master's degree, 30 credits or more from a university for the advanced class certificate; a bachelor's degree for the first class certificate; and an associate degree from a junior college for the second class certificate. The number of credits in subjects relating to the course and the professional qualification required to obtain a regular teacher certificate, are stipulated by laws and regulations in accordance to the type of certificate. For example, a first class certificate for lower secondary school teaching in Technology Education requires 20 credits of course-related subjects, 31 credits of subjects related to the teaching profession, and eight credits of subjects relating to the course or the teaching profession. Incidentally, five credits out of 31 credits related to the teaching profession are allocated to student teaching.

It was decreed necessary in 1998 to take a seven-day nursing care course at a special education school or a social welfare institution, in order for students who enter the university to acquire a regular teacher certificate for elementary school or lower secondary school. On the other hand, incumbent teachers who have already obtained a teacher certificate can obtain a higher class of certificate by acquiring prescribed credits after working at a school for a certain period of time with acceptable levels of competency.

A special certificate is granted to members of society with professional expertise or skill (such as computer-related workers) who have a bachelor's degree, to enable them to become employed as teachers. While regular teacher certificates are valid for life in all prefectures, the special certificate is valid for a period of not more than ten years and only in the prefecture where such a certificate is granted.

A temporary certificate is a certificate for an assistant teacher, which is granted only when a teacher with a regular teacher certificate cannot be employed. The temporary certificate is valid for three years in the prefecture where such certificate is granted.

CONCLUSION

National universities of Japan, which had been formerly and directly governed by the Ministry of Education, Culture, Sports, Science and Technology, were converted to individual national university corporations in April 2004. Although they will continue to be operated under the national budgets as in the past, competition among universities for research and teaching will be intensified, and independence in fundraising will be emphasized. On the other hand, the authority of the president or the direction of the administration in terms of university management is likely to dominate within a university, posing a concern to many instructors, in areas of budgets and human resources for the future.

In particular, because the technology education course is generally smaller in the number of instructors but needs more facilities, equipment and budgets than other subjects, the situation is much more serious. In addition, the declining child population as a whole in Japan is a cause of concern over shortages in qualitative and quantitative expansion of technology education as a specialty. It is not uncommon for other subject teachers to teach technology classes in small schools because there is no technology teacher. This trend also leads to a decrease in the number of technology teachers employed, and a further decrease in the number and morale of students who aim at Technology Education.

In future, the integration of national universities is demanded as part of recent national university reforms in Japan. As a result of voluntary integration, the number of national universities decreased from 99 in 2002 to 89 in 2004. In addition, the plan to integrate departments of education and teacher's colleges by region was realized and some departments and universities have already been integrated and are rationalizing roles shared among them. These situations mean that it will become impossible to carefully foster technology teachers by individual prefecture. It poses a serious issue not only to technology educators but also to all instructors in teacher's colleges/departments. We need to strongly publicise the importance and necessity of technology education as education for all.

REFERENCES

Aichi University of Education (2004). *Curriculum Guide Book for Students 2004*. Aichi: Aichi University of Education.

Ministry of Education, Science, Sports and Culture (2000). *Education in Japan 2000—A Graphic Presentation*. Tokyo: MESSC, pp.1–42.

Tokyo Shoseki (2002). *Guide Book for Technology Teachers in the Industrial Arts and Homemaking Education*. Tokyo: Tokyo Shoseki, pp.140–155.

Technology Teacher Education in New Zealand

Chapter 8

Alister Jones
University of Waikato, New Zealand

INTRODUCTION

The New Zealand compulsory schooling sector is typically made up of primary schools (years 1–6) or full primaries (years 1–8), intermediate/middle schools (years 7–8/7–9) and secondary schools (years 9–13). There are other schools that cross or include these boundaries but in terms of teacher education the years of schooling will be identified as primary, intermediate/middle and secondary. Children in New Zealand traditionally start school on their fifth birthday.

New Zealand has had a long history of technical education in the senior primary and secondary school (Burns, 1992). A national school system was introduced in New Zealand in 1877 and technical education was introduced in 1890 with metal and woodwork for boys and cooking, needlework and/or laundry for girls, being taught in the last two years of primary schooling (10–12 years old). At the same time, technical high schools were developed and these tended to channel working class children into manual and trade employment. After 1945, common core subjects such as metal, woodwork, cooking and sewing were introduced in all high schools for third and fourth form students (13–15 years).

During the 1970's and 1980's there were moves to include more design and the use of a broader range of materials. This saw the emergence of such subjects as workshop technology and graphics and design. During this time there were also attempts to break down the gender stereotype by permitting females and males to take all the technical/technology subjects in senior primary and junior high school. However, by senior high school these subjects tended to be gender specific (McKenzie, 1992). Also during the 1980's there was an increasing emphasis of technology in existing school subjects such as science (technology as applied science), social studies (technological determinism) and information technology (computers). Technology therefore as it has developed in past curricula encompassed a limited range of skills, processes and knowledge resulting from a narrow and gender specific perspective.

In 1990, a new government embarked on a project to revise the curriculum in primary and secondary schools, under the banner of 'the Achievement Initiative' (Ministry of Education, 1991). Many of these ideas were influenced by the curriculum reforms that were taking place in England and Wales from 1988-1991. The policies emphasized raising standards; levels of attainment and the notion of progression linked to accountability.

As part of the educational review process, a Ministerial Task Group Reviewing Science and Technology Education was set up jointly by the Minister of Education and the Minister of Research, Science and Technology, in June 1991, and which reported in 1992 (Ministry of Research, Science and Technology, 1992). Some of the recommendations from the task group concerning technology education included: a technology curriculum to be developed as an area in its own right; adequate teacher training and resourcing for technology education; technology curricula not being imported from overseas; the inclusiveness of technology education be emphasized, particularly, Maori input and the use of Maori language[1]. This report was endorsed by members of the business community, and by two Ministers of the Crown.

The New Zealand Curriculum Framework (Ministry of Education, 1993) provided an overarching framework for the development of curricula in New Zealand and defined seven broad essential learning areas rather than subject areas. The seven essential learning areas that describe in broad terms the knowledge and understanding that all students need to acquire are; health and well-being, the arts, social sciences, technology, science, mathematics, and language and languages. The New Zealand Curriculum Framework requires that all national curriculum statements in the essential learning areas specify clear learning outcomes against which students' achievements can be assessed. These learning outcomes or objectives are defined over eight progressive levels and grouped in a number of strands. This meant that a technology curriculum would define the broad area of learning and each school would base their individual curriculum on the national objectives.

---

[1] Maori are the indigenous people of New Zealand and under the Treaty of Waitangi (1840) were guaranteed exclusive and undisturbed rights in terms of preservation of land, fisheries, forest and language. Maori has been an official language in New Zealand (along with English) since 1987.

The general aims of technology education in *Technology in the New Zealand Curriculum* (Ministry of Education, 1995) are to develop:
- Technological knowledge and understanding,
- An understanding and awareness of the interrelationship between technology and society,
- Technological capability.

The three interrelated general aims provide a framework for developing expected learning outcomes, and make a valuable contribution to formulating a balanced curriculum for technology education. The technological areas through which these aims are realized include: materials technology; information and communication technology; electronics and control technology; biotechnology; structures and mechanisms; process and production technology; and food technology. Each technological area has its own technological knowledge and ways of undertaking technological activity. It is important therefore that students experience a range of technological areas and contexts to develop an understanding of technology and technological practice. Theories of learning also point to the fact that the more students can work in a number of contexts and areas, then the more likely they are to develop effective knowledge about technology and transfer this knowledge to other contexts and areas (Perkins and Salomon, 1989). The individual objectives of the technology curriculum over eight levels arise from the general aims of technology education.

Technological Knowledge

Students need to develop an understanding of the principles underlying technological developments such as aesthetics, efficiency, ergonomics, feedback, reliability and optimization. These knowledges and principles will be dependent on the technological area and context in which the students are working. The understanding of systems is essential in developing knowledge in technology. Students will also need to develop an understanding of the nature of technological practice and how this has similarities and differences in different technological communities of practice. It is important that students have an understanding of a range of technologies and the way in which they operate and function. An understanding of strategies for the communication, promotion, and evaluation of technological ideas and outcomes is integral. For example, five to six year old students within the level 1 group of objectives would be expected to understand the simple operation of some technologies, identify modifications, identify input and outputs in systems, examine how people carry out technological activities and how technological ideas are communicated, such as in advertisements. These objectives are undertaken in a number of different technological areas. Towards the end of high school, at levels 7–8, students would be expected within a range of technological areas to: explore and appraise the relationship between the use, function and operation of specific technologies; analyze technological principles; appraise complex systems in terms of interconnectivity within and between systems; critically evaluate a specific technological innovation in relation to the way innovations are communicated and promoted.

Technological Capability

Technological activity arises out of the identification of some human need or opportunity. Within the identification of needs and opportunities students will need to use a variety of techniques to determine consumer preferences. In technological activities, students should develop implementation and production strategies to realize technological solutions. Part of this will involve students in developing possible ideas that will lead to solutions and develop and use strategies to realize these ideas. Within this, students will need to manage time, resources, and people and produce the outcome that meets the identified needs and opportunities. Students should communicate their designs, plans and strategies and present their technological outcomes in appropriate forms. Part of this process is devising strategies for the communication and promotion of ideas and outcomes. Throughout the technological activity, students should continually reflect upon and evaluate the decisions they are making. Research indicates (Jones and Carr, 1993) that this is essential if students are to realize their technological outcomes. For example, technological capability for five to six year olds is in terms of identifying needs and opportunities, developing ideas and possible solutions, identifying necessary resources, and describing the process they went through. For the later years of high school, technological capability is represented as students using a variety of appropriate techniques to identify, clarify, and review needs and opportunities, taking into account consumers, the market and preferences; identifying task specifications, establishing constraints and considerations, applying relevant research to generate viable preliminary solutions and strategies, and testing, selecting, and modifying or adapting a viable solution; developing a critical path, managing resources effectively; producing a solution which meets the identified requirements, including production efficiency, quality assurance, commercial standards of performance, and health and safety standards; making informed decisions about and implementing strategies for the communication and promotion of ideas, decisions, and outcomes, with reference to community responses; critically analyzing and evaluating the strategies and outcomes, appraising the systems used to meet standards and criteria related to performance, aesthetics, and consumer demand. These general objectives are met across the different technological areas.

Technology and Society

Students should develop an understanding of the ways in which beliefs, values, and ethics promote or constrain technological development and influence attitudes towards technological development. Students should also develop an awareness and understanding of the impacts of technology on society and the environment. For example, senior high school students would be expected to: investigate and analyze ways in which beliefs, values, and ethics of individuals and groups promote and constrain technological developments in specific communities; analyze and critically evaluate the social and economic impacts of some significant technological developments in a variety of settings, debating viewpoints, and exploring options for the future. Technology and society is integral to technological activity so these objectives can be met within the context of a technological activity in the different technological areas.

Throughout the curriculum statement there are possible learning experiences and assessment examples for every objective in each technological area and which give advice to teachers about what should be introduced at the different levels. Progression is expressed in terms of increasing number of variables and the complexity of those variables rather than in terms of increasing complexity of solution or skills development. Progression may be more valuably conceptualized as a broadening and widening of the issues identified as essential to a technological activity (Jones and Moreland, 2003). Compared with countries such as England and Wales, design and the design process have not been mentioned as a separate achievement aim in the New Zealand technology curriculum. Design is a vital and inseparable part of technological capability and is therefore integrated throughout the curriculum. Different technology areas often have different views of designing and this can be taken into consideration if broader views of design are incorporated.

HISTORY

Since 1997 there has been a dramatic increase in the number of teacher education providers from an existing seven to over 30 in a country of just over four million. As highlighted in the previous section, the draft technology curriculum was released for comment in late 1993 and the final version was released late in 1995 after a period of national consultation. During this time, teacher education providers began considering how technology could fit into their programmes, particularly in the curriculum/subject studies areas. When the curriculum was implemented in February 1999, many of the existing pre-service teachers had to prepare to teach technology upon their graduation. Courses began to appear in teacher education from 1997 onwards. These however took a variety of forms, depending of how the teacher education provider viewed the curriculum area. For example, some perceived that technology was a separate curriculum area with its own knowledge base while others did not consider technology an area in its own right but rather a subset of other curriculum areas and accordingly inserted aspects of technology into existing curriculum courses.

As teacher education institutions began to grapple with the introduction of technology education courses into teacher programmes, the Ministry of Education was undertaking national professional development programmes. For the implementation of technology, the Ministry of Education provided approximately $NZ22 million for professional development purposes. These national programmes had been informed by two teacher professional development programmes developed under contract from the Ministry of Education by the Centre for Science and Technology Education Research (1995–1997). These initial programmes were the National Facilitator Training Programme, and the Technology Teacher Development Resource Package (for full details of these programmes and the evaluation see Jones and Compton, 1998 and Compton and Jones, 1998). These teacher development programmes took into account past national and international research in teacher development, as well as recent technology education baseline research carried out in New Zealand schools (Jones, Mather and Carr, 1995).

Findings from previous research led Jones and Compton (1998) to argue that technological knowledge and an understanding of technological practice must be combined with appropriate conceptualisations of technology and technology education. This is consistent with Shulman's argument (1987) that teachers need to develop content knowledge as well as pedagogical knowledge. Teacher's existing ideas of teaching and learning generally, and their needs, expectations, and classroom experiences in technology education specifically, are also factors that must be taken account of in technology teacher development programmes. Key features of the programmes were the importance of:

- Developing a robust concept of technology and technology education;
- Developing an understanding of technological practice in a variety of contexts;
- Participants developing technological knowledge in a number of technological areas;
- Participants developing technological skills in a number of technological areas;
- Developing an understanding of the way in which people's past experience both within and outside of education, impact on their conceptualizations about technology education;
- Developing an understanding of the way in which technology education can become a part of the school and classroom curriculum. This must be based on a sound pedagogy in keeping with the concept of technology education.

Features of and the materials from these professional development programmes formed the initial basis of technology education courses at the pre-service level. Although each institution developed its own approach to teacher education, this initial work formed the basis of many programmes.

At the same time as technology courses were being developed at the pre-service level, primary teacher education changed from four-year to three-year degrees. Traditionally there had been three-year diplomas and four-year Bachelor of Education degrees. Colleges of Education and other non-university teacher education providers were permitted to offer degrees, so a proposal for a three-year teaching degree was approved. When three year teaching degrees were approved, providers of four-year degree programmes began providing three-year teaching degrees. This change in length of time for teacher education impacted more on the amount of time devoted to curriculum areas than on other areas of teacher education. Technology education in most primary pre-service education areas received only 36 hours of time over three years.

OVERVIEW OF TECHNOLOGY TEACHER EDUCATION

Technology Education forms a part of all pre-service education for all primary school teachers. It is compulsory since technology is an essential learning area of the New Zealand Curriculum Framework, with most institutions teaching technology education after this learning area was fully implemented in 1999. However there are differences in emphasis depending on the structure of the institution. Teacher education takes a variety of forms within the Colleges of Education and Universities and there are a number of pathways.

Pre-Service Education

The Bachelor of Teaching is a three-year degree that prepares students for teaching in primary, intermediate or middle schools. This programme can be delivered at a distance through web-based media from a number of institutions. The first distance learning students graduated in 2000. The Graduate Diploma of Teaching (Primary) is a one-year programme for University graduates wishing to teach primary, intermediate or middle school students. For secondary school teacher education there are currently two pathways: a one-year Graduate Diploma of Teaching (Secondary) programme for graduates to teach in areas of their initial degree specialization; and the Bachelor of Teaching (Secondary), which is a conjoint degree programme. In this latter programme, students study for a degree in a relevant discipline as well as studying in the teacher education programme. There are also Bachelor of Teaching (Honours) programmes that enable students gain greater depth and specialisation in various aspects of teaching.

The Bachelor of Teaching (Primary) programme gives grounding in all aspects of the New Zealand Curriculum – the Arts, Music, Health and Physical Education, English, Environmental Education, Maori, Mathematics, Science, Social Studies and Technology. It also provides the opportunity to strengthen a curriculum area by pursuing an area of personal interest through elective papers. This three-year programme has four elements, which are taught over the three years: professional education (which is the degree major), practicum, curriculum, and subject studies electives (or a subject studies elective).

- In the first year of the degree these four elements have the following weighting;

- Professional Education: 1
- Practicum: 1
- Curriculum: 5
- Electives: 1
• In the second year the weighting is
  - Professional Education: 3
  - Practicum: 1
  - Curriculum: 2
  - Electives: 3
• In the third year the weighting is
  - Professional Education: 1
  - Practicum: 1
  - Curriculum: 1
  - Electives: 1

With all seven curriculum areas to cover as well as literacy, numeracy and environmental education, there are typically only 36 hours for technology education at the first year level. Although it is possible to study further in technology very few, if any teachers do so, because of the structure and nature of the programmes.

All Graduate Diploma programmes are one year courses and students can specialise in primary or secondary school education. The diploma structure has three approximately equal components of professional education, practicum and curriculum teaching subject. Although the normal entry requirement into the secondary programme is a degree or degree equivalent, those wanting to teach technology, who have trade related qualifications and 6000+ hours post qualification experience can be considered for entry into secondary teacher education programmes.

Masters Courses in Technology Education

The postgraduate course development has been significant, since the majority of teachers who are currently teaching technology were trained without any formal technology education. With technology being identified as one of the seven learning areas of the curriculum in 1991 there was some interest from teachers in undertaking a masters course. In 1992 the policy work for the new curriculum was undertaken, which included national consultation as well as the production of so called up-dates that informed schools of the policy development (Jones and Carr, 1993). The first course in technology education at the graduate level started in 1993 at the University of Waikato. With the release of the draft for technology studies in the New Zealand Curriculum in late 1993, there were a larger number of teachers studying the course in 1994. The structure of these courses will be discussed in comparison with those at the undergraduate level.

Since 1995 one other university has developed a masters technology education programme and a College of Education developed a Technology education masters course at the later date. Two Universities (Massey University and The University of Waikato) developed masters programmes in technology education. From the initial face to face classes taught during semesters, technology education is now also taught in distance mode, at summer school and more recently there has been the development of fully web-based courses which are asynchronous and which can be delivered internationally. As well as developing courses in technology education, that is, courses with content related to the nature of technology, learning in technology, technology curriculum and issues such as teacher development and assessment, other courses have been developed which emphasise the content area of technology such as technological knowledge, technological innovation, and technology and society. At this graduate level, the courses follow a traditional design, which assists practicing teachers develop the knowledge to teach technology in the classroom.

STRUCTURE OF TECHNOLOGY TEACHER EDUCATION

This section builds on the previous section, providing examples of the different programmes including the structure, approaches and activities. Approaches will include the emphasis placed on theory and teaching and learning in the technology courses, as well as technological practice and knowledge. Ways of teaching technology will be highlighted, as well as ways in which students' knowledge of and about technology are developed. Curriculum approaches at the school level and in the classroom will be discussed, including national curriculum development, Colleges of Education and Universities have similar course structures and are often moderated between each other to maintain standards. However with the proliferation of teacher education providers this is not necessarily true for other providers. All three-year programmes have four elements taught over the three years. These are professional education (which is the degree major), practicum, curriculum, and subject studies elective. Technology is to be studied as a one of the seven curriculum areas. In the first year there is a 5:3 ratio in terms of curriculum to other areas whereas in the second year the ratio is 2:7 and in the final year there are no curriculum studies.

Primary and Middle School Teacher Education

As highlighted earlier, technology is a compulsory curriculum area of study in the pre-service teacher education programme. Issues related to general education are studied in the professional studies components of the degree programme. The pre-service courses in technology education in the Colleges of Education and Universities have a structure that is reflective of the early research in technology education in professional development. Pre-service courses are part of the curriculum papers that are undertaken at the first year level of the teacher education degree programme. From the information available the course structure generally consists of three general aims. These are: developing an understanding of technology and technology education, developing an in-depth understanding of the technology curriculum, and an understanding of planning, teaching and assessing technology in the classroom.

The general structure is firstly to explore concepts of technology and technology education. In the course material, students are introduced to different design concepts as well as the history of design and technology. Societal and values aspects of technology are highlighted for the students. Emphasis is placed on design and the process of undertaking a technological project to develop an understanding both of technology and also an awareness of issues that children in schools might have to deal with as they learn technology.

Through engagement in technology, students are also made aware of issues that they may face in the classroom, including methods for teaching technology, such as problem solving, group work, and resources. New Zealand has produced a number of resources for teachers and these are introduced to the students throughout the course. A television and video series (*Know How 2*, Ministry of Education, 1997) developed in 1997, provides insights into technology and technological practice. This education tool continues to be used in teacher education as an illustration for undertaking technological development. Early in the courses, students begin to develop their own technological projects.

The New Zealand Technology Curriculum has seven identified technological areas that are taught in schools. In the short time that is available for the compulsory curriculum areas, the emphasis in the courses has been to develop students' understanding of the nature of technology to create awareness of the characteristics of different technological areas. Of the seven technological areas, teaching time constraints result in only the areas of Structures and Mechanisms and Materials usually being highlighted, and these are also the most commonly taught areas in primary schools. Students are introduced to technical aspects, such as health and safety, drawing and construction techniques.

For the curriculum, students are introduced to its philosophy, strands and achievement objectives as well as examples of unit plans developed by other teachers to show the link between the curriculum and lesson planning. Students are required to connect these to the classroom experience. *Know How 2* also provides examples of classroom practice in technology as well as other written material.

Significant emphasis is placed on planning for technology in the classroom, considering scenarios through to curriculum objectives and then to specific learning outcomes. During this time the students are also introduced to ways of managing technology in the primary classroom. Formative and summative assessment is another significant area that is introduced to the students.

Secondary Teacher Education

Graduate Diplomas are one year programmes in which students can specialise in primary or secondary school education. The diploma structure has approximately three equal components of professional education, practicum and curriculum teaching subject. Secondary school pre-service courses in technology education assume that the students will have some prior content understanding and experience in technology. This understanding and experience can be gained through either an undergraduate degree or, as in an increasing number of cases, a university bachelor qualification and work experience in the technology sector. Some of the students may have come from an advanced trade or technician background. In secondary education there are also conjoint degrees where students can study at the undergraduate level in a technology discipline and jointly study in a teacher education programme. However, all pre-service students in the secondary technology programme undertake a course in the technology curriculum area. As highlighted earlier, the teacher education courses are split between professional studies, practicum and curriculum studies.

Emphasis in secondary technology education curriculum courses is on assisting pre-service teachers to translate their curriculum area content knowledge into planning, teaching and assessment strategies. Time is therefore spent on understanding the curriculum, its philosophy and intent, the different technological areas and the strands. Associated with this is an emphasis on student learning in technology prior to an introduction to planning, teaching, learning and assessing in technology. With the introduction of a new schools' exit qualification framework, increased emphasis has been given to understanding the NCEA (National Certificate in Educational Achievement). NCEA is undertaken in years 11, 12 and 13. Therefore in the secondary teacher programme, students need to develop a thorough understanding of the national certificate, as it is achievement based, being assessed throughout the year internally as well as by external assessment at the end of the year. A significant part of the secondary programme is the development of classroom teaching plans and strategies that can be trialled during the practicum section of the year.

General educational theory and practice in this course are undertaken in the professional studies part of the courses. Professional studies are approximately one-third of the course. This covers studies including general classroom management, educational theory, assessment and reporting issues, general pedagogical strategies, and coping with student diversity.

Graduate Education

When technology education was being introduced into New Zealand schools most of the teachers who would have responsibility for teaching this new curriculum area were already experienced classroom teachers. As highlighted in the previous section, graduate courses become an important part of the teacher education landscape in technology education, more so than in other curriculum areas. Although they reflect the interests of the developers in the various institutions, the graduate courses all tend to have much the same structure or intent. There have been two types of courses developed, one that explores technology education and the other in terms of the content area of technology.

Technology education graduate courses place considerable emphasis on the nature of technology and technology education. Technology education examines for example, the history and philosophy of technology, the nature of technological knowledge and technological development in society. The courses also examine learning theories associated with learning in technology and in particular, current research in learning in technology, such as technological problem solving, approaches to design and student conceptual understanding. An analysis of national and international technology curriculums is usually undertaken, as well as a study of the historical development of technology education. Graduate courses in technology education also consider such issues as professional development, gender and ethnicity, and assessment.

Other graduate content area courses in technology are much more varied and aligned to the providers' expertise. These courses are designed to develop the content areas associated with technology such as technology innovation and development, designing, manufacture, electronics and food technology, and often taught by university staff from technology or engineering departments.

AN EXAMPLE: UNIVERSITY OF WAIKATO

This was one of the first programmes established and was a direct result of early work in technology education research and curriculum development that had been undertaken at the University of Waikato. Course philosophy, structure, approaches and outcomes will be highlighted for both primary and secondary programmes. The one-year primary graduate programme will also be discussed.

The pre-service paper in technology education was first introduced in 1998 and is a 36-hour compulsory curriculum paper. The paper has three objectives:

- Examining and developing student ideas of technology and technology education. This is achieved through exploration of students' views of technology through discussion around examples and activities, working towards a shared view of technology consistent with the curriculum;
- Understanding the technology curriculum. This is achieved through examining the curriculum aims and intentions.

- Implementing the curriculum in the classroom. The student explores different technological areas and classroom planning.

This paper aims to achieve the above objectives by providing a range of authentic technological activities designed to help the student understand technology and its practice. In addition to a range of practical workshops covering a variety of technological areas, students undertake a project in which they work in teams to plan, implement and evaluate possible solutions to identified problems, needs or opportunities. These experiences and activities provide an important foundation from which to engage with the later part of the paper that examines the technology curriculum document as well as planning for its successful implementation in the primary classroom.

A central part of the paper is groupwork through a technology process. The purpose of this is to provide students with practical, first-hand experience of working through a technological project; to help them understand technological practice and assist them in developing authentic technology education programmes. In planning, managing, implementing and evaluating the project, students include a design brief, feasibility report and detailed specifications as parts of a project plan. They establish project milestones, project team responsibilities and consider issues of quality control and the management of human, time, physical and monetary resources. Their submitted portfolio records the various aspects of planning, managing, implementing and evaluating undertaken during the project and provides a record of the developmental process of the project as well as presenting the project outcome.

There are strong indications that undertaking this type of sustained project is very beneficial to students developing an understanding of technology and how they might think about teaching technology in the classroom. Students currently undertaking courses will not have studied technology before and will not have had any experience in technology. The vast majority of students have 36 hours of technology before becoming teachers. This course is delivered in an on-campus environment as well as being offered on-line. In both modes, students carry out a group technological activity. For the on-line mode, students working initially in groups during an on-campus week and then use 'chat' facilities to continue to develop their ideas and products. Students living in the same region may arrange to meet.

The technology education course is slightly different from courses such as science education. Whereas in science, students spend time working with children in schools, the coordinators of the technology course chose not to do this so that more time could be spent on developing students' knowledge of technology and technology education.

In the graduate one-year programme, the aims and length of the course are similar but technology education tends to be taught in a more academically attuned manner and the students do not necessarily do a project. More time and detail are spent on content and simple equipment that students may use in their teaching. Greater time is also spent on planning and understanding of technological areas as there is only one year to prepare them for planning and teaching.

TEACHER CERTIFICATION

To become a registered teacher in New Zealand an application is required to be lodged with the New Zealand Teachers' Council. It is normally expected that to become a registered teacher the candidate will have a recognised teaching qualification and have taught for at least two years with professional support and appraisal. The initial teaching qualification must also be recognised by the New Zealand Teachers' Council. The Teachers' Council moderates all teacher education providers' courses and programmes. A panel moderates the programmes with a chairperson appointed by the Teachers' Council.

The professional development component of the registration process can be seen in terms of:

- Continuing education in teaching or in the teaching subjects through tertiary courses, in-service courses, refresher courses, workshops as well as seminars;
- Research based activities involving a school or classroom based research, such as developing a new programme;
- Professional activities through, for example, active participation teacher subject. This professional development component is undertaken during the first two years of beginning teaching. Courses are also run by Teacher Support Services (in-service providers funded by a purchase agreement with the Ministry of Education) for beginning teachers. This is part of the government support for beginning teachers.

There are professional standards set out for both primary and secondary school teachers in terms of beginning teachers, full registered teachers and experienced teachers, which are usually linked to levels of pay. The dimensions are:
- Professional knowledge – curriculum, current issues, learning and assessment theory;
- Teaching techniques – planning and preparation, teaching and learning strategies, assessment/reporting, use of resources and technology;
- Motivation of students – engagement in learning, expectations that value and promote learning;
- Classroom management – student behaviour, physical environment, respect and understanding;
- Communication – student, colleagues, family/whanau;
- Support for and co-operation with colleagues;
- Contribution to wider school activities.

Progression along these dimensions is highlighted and can be used to both assist and monitor a teacher from the beginning through to registration. Although registration normally takes two years, for some teachers it can take up to five years.

CONCLUSION

Technology Education in New Zealand has become well established in the rhetoric and practice of the classroom. However there is a range of both practice and time spent on technology education in the classroom (Jones, Harlow and Cowie, 2004). Technology is being squeezed in a curriculum that is being influenced strongly by literacy and numeracy initiatives in the primary school and new qualifications in the secondary school. In terms of teacher education there has also been the decrease from four to three year initial teacher education degree programmes at the primary/intermediate school level. Very little time is available to prepare year 1–8 teachers in curriculum areas such as technology. Technology education courses at the pre-service level introduce students to technology as very few of the students obtained any background in technology while they were studying at secondary school. At the same time technology education courses are also required to provide the students with appropriate teaching strategies. However, this does not imply in any way that New Zealand teachers are not well prepared for the classroom generally or in areas such as literacy and numeracy. Research on teachers introducing technology in their classrooms would show that these teachers had very high levels of pedagogical knowledge and were highly effective practitioners (Jones and Moreland, 2004). It has also been shown that these teachers can become highly effective technology teachers with appropriate teacher professional development. At the secondary school level the main issue is to encourage more people from the technology sector to become technology educators in secondary schools.

Overall, technology is becoming part of the school curriculum and teacher education at the pre-service level is beginning to contribute to this. However, there is still a significant way to go until technology is seen in the same light as other well-established subjects such as science and social studies.

REFERENCES

Burns, J., (1992) Technology - What is it, and what do our students think of it? *The NZ Principal,* 6, 22-25.

Compton, V., and Jones, A., (1998) Reflecting on teacher development in technology education: Implications for future programmes. *International Journal of Technology and Design Education.* 8, 2, 151–166

Jones, A., and Carr, M., (1993) *Student technological capability.* Vol 2: Hamilton: Centre for Science and Mathematics Education Research, University of Waikato.

Jones, A., and Compton, V., (1998) Towards a model for teacher development in technology education: from research to practice. *International Journal of Technology and Design Education.* Vol 8, 1, 51–65.

Jones, A., Harlow, A., and Cowie, B. (2004) New Zealand teachers' experiences in implementing the technology curriculum. *International Journal of Technology and Design Education.* Vol 14, 2, 101–119.

Jones, A., Mather V., and Carr, M., (1995) *Issues in the Practice of Technology Education.* Vol 3: Centre for Science and Mathematics Education Research, University of Waikato.

Jones, A., and Moreland, J. (2003) Developing classroom-focused research in technology education. *Canadian Journal of Science, Mathematics and Technology Education* 6, 51–66.

Jones, A., and Moreland, J. (2004) Enhancing practicing primary school teachers' pedagogical content knowledge in technology. *International Journal of Technology and Design Education.* Vol 14, 2, 121–140.

McKenzie, D., (1992) The Technical Curriculum: Second Class Knowledge, In McCulloch G., (Ed) *The School Curriculum in New Zealand: History, Theory, Policy and Practice.* Palmerston North: Dunmore Press.

Ministry of Education, (1991) The Achievement Initiative. *Education Gazette,* 70, 7, 1–2, 16 April.

Ministry of Education, (1993) *New Zealand Curriculum Framework.* Wellington: Learning Media

Ministry of Education, (1995) *Technology in the New Zealand Curriculum.* Wellington: Learning Media

Ministry of Education (1997) *Towards Teaching Technology: Know How 2.* Wellington, Learning Media

Ministry of Research, Science and Technology (1992) *Charting the Course: the report of the Ministerial task group into Science and Technology Education.* Wellington: Government Printer

Perkins, D N and Salomon G. (1989) Are Cognitive skills context bound? *Educational Researcher,* 18, 1, 16–25.

Shulman, L. S. (1987) Knowledge and Teaching: Foundations of the New Reform. *Harvard Educational Review* Vol. 57, 1, 1–22.

ACKNOWLEDGMENT

I am grateful to Dr Michael Forret for his helpful comments in the development of this chapter.

Technology Teacher Education in Russia

Chapter 9

Margarita Pavlova
Griffith University, Australia

INTRODUCTION

In 1993, Technology Education replaced the subject Labour Training, which had previously occupied a significant position in the Soviet curriculum. Technology education was introduced as a compulsory learning area in Russian state schools (where the vast majority of students study), with 808 hours allocated over the period from Years 1 (Y1) to 11 (Y11). Four historical features influenced the nature of the new learning area.

Firstly, Technology Education continues to be a part of general education. After the 1917 revolution, technical and practical subjects were articulated as a part of general education. Labour Training remained a separate subject in the school curricula for most of the Soviet period and was compulsory for students of all grades.

Secondly, besides the ideological underpinning (everyone had to be socialised into the proletariat culture) a strong engineering tradition in understanding technology influenced the establishment of the polytechnic principle, a teaching paradigm based on teaching the scientific principles that underline manufacturing processes, and training students in practical skills with a variety of tools and equipment.

Thirdly, highly structured knowledge was 'transferred' to students. It was a reflection of the pedagogical assumption that a person cannot be involved in the activity without the prior knowledge that is related to it. So, the belief was that theoretical knowledge should be learnt and then applied through practical activity. In a typical Labour Training lesson, 25% of time was allocated to the theory at the beginning of the lesson and the remaining 75% was allocated to the practical activity. Students made identical objects, using the provided instructions.

Finally, the technology education curriculum was introduced in a very centralised way. Confidence in the state's strong role in the education process goes back to the revolution of 1917. There was a strong belief that the transmission of a universal curriculum was a pathway to liberty, equality and fraternity. Central control of the school curriculum has been the main managing principle for decades.

Currently, although power is divided between federal, regional and school levels, and the development of curricula for primary and secondary schools is their shared responsibility, the federal component of the curriculum for technology education constitutes a major part of the curriculum. It is specified through the *Standards*.

The nature of the first Standards for technology education (Lednev, Nikandrov, Lazutova, 1998) still provides the basis for school programs in Russia. It remains mainly unchanged, compared to Labour Training, and incorporates all the features described above. As a knowledge-based paradigm for education, it shaped students' understanding of technology. Technology is defined as – a science [body of knowledge] regarding the transforming and using of materials, energy and information for the purpose and interest of man – (Lednev, Nikandrov, Lazutova, 1998, p.247). Aims of technology education are to:

- develop students according to the *polytechnical principle*, to acquaint them with modern and prospective technologies of processing materials, energy and information via the application of knowledge in the areas of economics, ecology and enterprise;
- develop general working skills;
- stimulate the creative and aesthetic development of students;
- acquire life-needed skills and practices, including the culture of appropriate behaviour and non-conflict communication in the process of work;
- provide students with the possibilities of self-learning and studying for the professions, and the acquisition of work experience which could be the basis for career orientation (Lednev, Nikandrov, Lazutova, 1998, p.248).

Reference to the polytechnical principle indicates that the applied science paradigm is still in force: objective knowledge provided by science can be used for developing technological innovations. The educational approach to transfer the relevant knowledge is still in place. There is no acknowledgement that students can construct their knowledge through practical activities. The role of content is extremely important and is specified in detail in the Standards. The compulsory minimum content for the Technology learning area is identified for both city and rural schools (with differing contents for boys and girls).

The tradition of having a separate curriculum for boys and girls is hidden by the statement advising it is important to cater for different students' interests. Thus, students have to choose one of the following subjects according to their interest:
- technic (processing resistant materials and electronics), or
- culture of the home (house-keeping work).

Modernisation of Russian education began in 2001. The Strategy of Modernisation (The Ministry of Education of the Russian Federation, National Fund for Personal Training, 2001) is aimed at overcoming the historical assumptions discussed above. School reform is being undertaken in a variety of areas such as: expanding general schooling to twelve years and reforming the assessment rationale by combining school graduate exams and university entrance exams. However, the major aspect of reform is the modernisation of the school curriculum content. Goals of modernisation for Russian education systems are ambitious: to achieve modernisation of the whole country/state by means of content and structural renewal of education. The main emphasis in the modernisation of school content is on development of the 'cultural' person who would potentially be able to solve problems in different fields. Thus, the potential ability to solve problems is considered as a principle goal of education.

A major claim proposed that the content of education should be structured on the basis of different spheres of human activities that constitute its culture (The Ministry of Education of the Russian Federation, National Fund for Personal Training, 2001). Students need to develop competencies in the following areas:
- cognitive activities (based on methods of mastering strategies for acquiring knowledge from different sources of information);
- civil-social activities (roles of the citizen, voter, consumer);
- socio-working activities (including the ability to analyse the situation within the labour market, to evaluate personal professional abilities, and to establish an orientation to the norms and ethics of labour relationship);
- the household sphere (including aspects of health and family well-being);
- culture-leisure activities (including options for using non-work time that culturally and spiritually enrich the person).

A process of designing new Standards focussing on the development of these competencies commenced in 2002. The rationale for developing these Standards states that they should include a re-orientation from the content-based approach to the activity-based approach in teaching and learning. Thus, the outcomes of the learning should be formulated through the patterns of activities, and fulfilment of the task by the students will demonstrate their achievements.

A draft of the second Standards for technology education was published in 2003, with the aim of implementing the Standards in two stages, starting in 2006. In accordance with the Strategy of Modernisation, the aims of technology education, as stated in a draft of the second set of Standards (The Federal Component of the State Standards for the General Education, draft, 2003) are less oriented towards knowledge acquisition, and more oriented towards the personal development of students: developing inquiring minds, technical thinking, spatial imagination, intellectual, creative, communicative and management skills, self-directed involvement in activities, mastering a technological culture, as well as orienting a pedagogy towards diverse activities aimed at creating personally and socially useful products. Also, the concept of projecting (design) has been introduced. The Standards are less directly related to a particular type of work after school. They are aimed at preparing students for life and work in general.

Another difference with the 'old' Compulsory Minimum content of technology education is that the new Standards consist of two components: a *general technology* component and a *specialised* component. General technology is compulsory for all students while specialised education consists of several options from which the students choose one. The general technology component includes the following content: main technological concepts and types of activity, basis of transformative and design activities, technological and consumption culture and professional orientation (career guidance). Professional orientation should be integrated with one of the specialised components: 'directions' (technology of hand and machine manipulation with resistant materials, artistic development of materials; technology of textile and food, culture of the house; technology of agriculture – for rural schools) or 'areas' (the practical activities of humans: manufacturing industry, economics, education, medicine, building industry, transport, Information Technology (IT), applied art and craft, office and secretarial studies, horticulture, animal husbandry, and service industries). The argument is that this structure provides the possibility for students to study the basis of technological culture as an element of the general culture within the specialised classes such as social-humanities, humanities-philology, science, physics and mathematics. The specialisation is known as *profiles*. It is believed that this approach will better suit students' interests.

Thus, technology education in Russia is in the process of transition from one set of Standards to another, and one which presents a different pedagogical paradigm. The number of hours for technology education in the curriculum is under discussion at the moment. It is proposed that in Y8 and Y9 technology education will be reduced from two hours to one hour per week.

EDUCATION SYSTEM IN RUSSIA

Compulsory education in Russia comprises nine years of schooling. Children commence school at the age of six, attending primary school (Y1 – Y4) for four years. Then they move to secondary school (Y5 – Y11) which consists of two components: main secondary school Y5 – Y9, which is compulsory, and Y10 – Y11, which is non-compulsory. After Y9, students decide between leaving the main secondary school, going to work, going to study at different types of vocational schools or to stay at school for the remaining two years. The state guarantees free education for all until the end of secondary education. Access to further levels of education is competitive.

Vocational education in Russia is structured at several levels and refers to any post-general education that provides training for a career. Two of the levels of vocational education are: first level after Y 9 for two or three years, and second level after Y 11 for one or two years. According to UNESCO's international standards classification, the second level of vocational education is equivalent to practice-oriented higher education or pre-university higher education (The program of vocational education development, 2004). It can start after Y9 (11% of graduates), or after Y11 (23% of graduates), and the study can last from one year ten months to four years ten months, depending on the specialisation (Review of vocational education development in 2001–2002, 2004).

Figure 1. Typical structure of primary and secondary vocational education.

Primary vocational education					Year	Secondary vocational education		
Age						Lycee	Polytechnic	College
19	Vocational schools	Lycee	Polytechnic	College		1–2 years		
18					11	Complete secondary school		
17		2–3 years			10			
16		Compulsory school			9			
6					1			

The major difference between the above two levels is the depth of study, which usually depends on the specialisation, for example, designers would be trained at the second level and tailors at the first level of vocational education. The number of students who are enrolled in lower level vocational education has declined during the last ten years by 17% (Vocational education, 2003). The number of students who are enrolled in the second level of vocational education has increased by 27% during the last five years (Russian Education by 2001: Analytical overview, 2004). This demonstrates the trend towards the higher level of education and re-establishment of the role of education in society in terms of career opportunities.

The third level of vocational education is higher education, with three types of institutions involved: universities, academies and institutes. These institutions (VUZ) also provide a post-university vocational education at the level of *Kandidat Nauk* degree (research degree that is accepted as PhD in other countries), then PhD degree and then postdoctoral studies as well as in-service training for higher education staff. The content of educational programs, number of teaching hours and requirements for graduates' qualities are specified by the state educational standards for higher education. Students can study full-time, part-time or externally (a mixture of remote study and short on-campus intensive sessions).

After a decline in student numbers during the mid-1990s (in 1995 there were 189 students per 10,000 people), student numbers are now growing. In 2000 the number of students per 10,000 people were 327 (Statistics of Russian Education, 2004). Currently, the system of higher education in Russia includes 607 state and 358 non-state higher education institutions, in which 4.7 million students are studying. Academic staff of the state universities, academies and institutes (VUZ) number around 265 000. Higher education prepares students for more than 350 specialisations, and 82 institutions train technology education teachers (Statistics of Russian Education, 2004).

HISTORY

The origin of technology education is vocational education (VET). VET started in Russia in the 18th century, during the time of Peter the Great, when the emphasis was put on civic education in opposition to religious education. In 1721 Tatischev, a philosopher, supervised the establishment of experimental vocational schools that were designed to develop a non-religious world outlook, and to educate students to be successful in life (Piskunov, 2001). Several vocational schools were established at that time, including; mining, medical and engineering. A number of teachers for these schools were recruited from abroad. Peter the Great had decided to change national *imaginaries* of that time by introducing Western trade training. The term imaginaries signifies the conceptualisation of the process of socialisation into national culture. Russians greatly opposed this process as it was contrary to the social, political and religious views of that time. Since then, the development of vocational education in Russia has developed in a way that reflects the complex relationships between local and external, social and economic development and VET. Until the 19th century, vocational 'teachers' were craftsman, who were teaching their skills to apprentices by involving them in practical activities.

At the end of the 19th century there were a number of experiments designed to include vocational training into education. The discourse was about the role of Labour Training in students' general education. Behterev, for example, argued that only 'socio-labour' up-bringing (value development) could provide the potential to overcome tensions between the demands of the 'person of society' and her individual interests and needs (Piskunov, 2001). From this time onwards Labour Training became an important part of general education.

Preparation of teachers for Labour Training has gone through several stages. As the training system in Russia has always been centralised, the history of technology teacher training can be explored through the history of one of the oldest pedagogical institutes in Russia – Herzen Pedagogical University, in St Petersburg. This training centre, that was established in 1797, grew out of the *Up-bringing House* for orphans and poor children. A number of factories were opened by the Tsarist government to provide work-places and income for the institution. Working at the factories, students learnt a particular profession and earned money to support themselves. By the end of the 19th century a vocational school had been established at the Up-bringing House that gave primary education and professional training for poor girls. Those graduates who successfully passed an exam on pedagogy after three years of work at the factory became teachers in handcrafts in this vocational school. That was the first model of teacher training for practical areas.

The middle of the 1920s presented new challenges for teacher training. Industrialisation of the country after the revolution of 1917 required everyone to study the basis of manufacturing industry. All students (including the humanities faculties) at Herzen (a former Up-bringing House) studied such courses as 'Introduction into modern technology' and 'Basis of manufacturing'. Students were also involved in industry placement practicums, where they learnt skills of a joiner, locksmith, blacksmith, and electrical engineer. They also learnt about the mechanical development of materials and had excursions to factories and manufacturing plants. As a result of this training, all teachers were able to provide technology education for students.

In response to the increasing demand for qualified workers, a new type of school was established: the factory–industry's seven-year schools. In 1930, a faculty that was training teachers for these schools was opened at Herzen. In the 1950s, during the post-war period, the issue of training of highly qualified workers was again raised, and that provided the impetus for changes in both school and university programs.

During 1957, the Physics-Mathematics faculty started training teachers with qualifications in what was described as *Physics and the basis of production and technical drawing*. The big difference between the quality of general and vocational school graduates and the requirement of socio-technological progress of the country became highly visible in the mid-1970s. Thus, the demand for qualified Labour Training school teachers and teachers for vocational schools was again raised. To meet this demand, a new department within the Physics faculty at Herzen was opened in 1978. The General Technology Department started to train Labour Training teachers within the higher education degree.

In 1979 this department was transformed into the faculty that was later renamed the Industry-Pedagogy faculty. In 1984, an additional specialisation was added in career guidance. In 1995 the faculty started training students as teachers of Technology and Enterprise. As a result of this change, in 1996 the faculty was renamed the Faculty of *Technology and Enterprise.*

This history reflects the process of establishing and developing technology education faculties that train technology teachers all over Russia. Now there are more than 80 faculties throughout the country.

OVERVIEW OF TECHNOLOGY TEACHER TRAINING

Technology Teacher Training

Technology teacher training programs have been undergoing changes in accordance with the whole system of higher education in Russia throughout the last decade. The number of technology education graduates has risen from 4712 in 1998 to 6227 in 2002. This can be compared with the total number of graduates in the two periods: 470,589 in 1998 and 840,405 in 2002 (Russian Education: Federal Portal, 2004).

Technology teacher training for secondary school is conducted at the level of higher education (the third level of vocational education in Russia). Technology teacher training for primary school can be located at both the second and third levels of vocational education and is undertaken within the general framework for training primary teachers.

The agreement on the common higher education sector across Europe (signed in Bologna in 1999) started the process of integration of Russian higher education into the European system of higher education. Bachelors and Masters degrees are being gradually introduced in all areas of studies. Now graduates can have the following qualifications: Bachelor Degree, specialist (for example, technology education teachers, economist, engineer) or Masters Degree. The structure of study programs can be continuous (as a five to six year program) or based on the levels. The number of students entering the first year of higher education is regulated by the federal bodies of executive power which are partly in charge of the institution's budget.

MODERNISATION OF HIGHER EDUCATION

The modernisation of the Russian education system discussed above also led to innovations in pedagogical education. At the national level, a committee to develop an action plan concerned with the modernisation of pedagogical education was established. The aim of the program is to provide sustainable and effective functioning of the system and its

development. Currently, the following tasks related to higher education are being addressed:
- Renewal of the list of directions and specialisations (qualifications) for higher education;
- Developing a system of forecasting the changes in demands for pedagogical staff;
- Monitoring the education system and the process of its modernisation;
- Conducting research to identify the priority areas for research in education;
- Supporting fundamental and applied research, such as in the established scientific (research) schools (The Ministry of Education of the Russian Federation, 2003a, pp.21-22).

Issues highlighted in the annual report on the Russian education system in 2002 (The Ministry of Education of the Russian Federation, 2003a) will have the effect on technology teacher training of a uniform approach to the content of teaching, experimentation with distance teaching, and use of credit points. Currently, of 82 institutions training technology teachers, only a few are heavily involved in research.

STRUCTURE OF TECHNOLOGY TEACHER TRAINING

Technology teacher training in Russia is regulated by the State Educational Standards for higher vocational education (The Ministry of Education of the Russian Federation, 2000). The Standard for Technology Teacher Training (graduates receive qualification 030600) which regulates the training of teachers of Technology and Enterprise, was put into force in 2000 and specified five years of post-secondary full-time studies. The Standard also describes the qualities of the technology teacher graduate and educational programs. They include the disciplines regulated at state and university levels, type of disciplines elected by students and special facultative (non-compulsory) disciplines. Elective courses in each cycle should supplement disciplines specified in the federal component of the cycle. The program should include the following cycle of disciplines:
- Humanities and socio-economic,
- Mathematics and science,

- General professional,
- Those specific for the particular learning area,
- Facultatives.

Training of technology teachers is undertaken through one of two approaches: a five-year continuous program or a multi-level program.

1. Five Year Continuous Program

The Standards established a duration of 260 weeks for the training program, comprising the following:

Table 1.
Structure of the training program.

Theoretical studies, including students' research work Practical studies, including labs	156 weeks
Exam sessions	27 weeks
Practicum (school and industry placement)	Not less than 28 weeks
Final state attestation, including the preparation and defence of the graduation project	8 weeks
Holidays (including 8 weeks after-graduation holiday)	38 weeks

The Standard that regulates training of technology education teachers also specifies:

- Maximum amount of study hours is 54 hours per week, including university classes and personal studies, but not including facultative studies (facultative studies are defined as non-compulsory courses provided by the university);
- University studies should not be longer than 27 hours of classes per week. Physical training and facultative lessons are not included in these hours;
- Holidays should be seven to ten weeks during the year, with a minimum two weeks during the winter break.

Standards specify the number of hours for each discipline to be taught. They are summarised in Table 2.

Technology Teacher Education in Russia

Table 2. Disciplines specified in the Standard.

Name of the discipline	Number of hours
Humanities and socio-economic disciplines	**1500**
Federal component	1050
Foreign language	340
Physical culture	408
Russian history	
Culturelogy	
Politology	
Law	
Russian language and the culture of speech	
Sociology	
Philosophy	
Economics	
National-regional component	225
Elective disciplines and courses identified by the University	225
Mathematics and science	**1000**
Federal component	850
Mathematics	334
Informatics	72
Physics	300
Chemistry	72
Ecology	72
National-regional component	150
General professional disciplines	**1600**
Federal component	1280
Psychology	280
Pedagogy	300
Basis of social pedagogy and psychology	72
Theory and methodology of teaching technology and enterprise	340
Different age anatomy, physiology and hygiene	72
Basis of medical knowledge	72
Safety through life	72
Technical and audio-visual means of teaching	72
National-regional component (University)	160
Elective disciplines and courses identified by the University	160
General technical disciplines	**4334**
Federal component	
Applied mechanics	300
Theory of machines	300
Technological disciplines	380
Information technologies	150
Electro-radio-technology	300
Graphics	300
Basis of the enterprise activities	280
Basis of creative-constructive activity	150
Technology practice	414
Specialisation disciplines (Specialisation)	**900**
National-regional component (University)	430
Elective disciplines and courses identified by the University	430
Facultatives	450
Military training	450
Total	**8884**

Flexibility of the program is limited to 5% variation of the number of hours specified in the Standards. Every teacher training program must be approved by the Learning–Teaching Methodology Union committee (UMO in Russian abbreviation). UMO is a state-public union in the system of higher education in Russia. It was established with the aim of coordinating the effort of academic staff, industry and representatives of other institutions to maintain the quality and development of the content of higher education, as well as to forecast future directions and provide support for the process of training students. Among the aims of UMO is: development of drafts of state Education Standards; provision of examples of study programs and curricula for higher education; approval of the list of directions and specialisations for higher education; review of manuscripts of books and textbooks prepared for the approval of UMO and the Ministry of Education (UMOa, 2004).

The draft of the second Standards for training technology teachers specifies the possibility of having different specialisations within one qualification. For example, within the Technology and Enterprise qualification, the specialisation is a 900 hour program (see Table 2). The specialisation should constitute a coherent program that provides the opportunity for students to develop knowledge and skills that help them work in a rapidly changing education environment, as well as their gaining a systematic training in a particular area.

Presently, there are more than 28 specialisations within the qualification for a teacher of Technology and Enterprise. Some of these are: technic and technical creativity, culture of the house and applied art, peasant house and the family, building and maintenance of the individual dwelling, textile technology, food technology, design of household and industrial products, graphics and design, high-tech, applied economy, technology of agriculture, professional orientation (career guidance), foreign language in vocational education, and labour training. The number of specialisations offered in each institution varies, with some universities offering one while other institutions offer up to a maximum, in practice, of ten specialisations (UMOb, 2004).

Specialisation provides flexibility in the training system, allowing it to respond to the demands of education practice. Specialisations can be quickly approved through UMO, unlike the new educational Standards that must be established and approved by the government.

2. Multi-Level Program

Multi-levelness of education is a new model for higher education with the aim of integrating Russian higher education into the European system. In 1994, a decree of the Russian government established a general approach for the structure of the multi-level higher education sector (Russian Government, 1994). The first level of higher education is a non-complete higher education. Students can receive a diploma after two years of studies. The second level is a four-year Bachelors program. For education qualifications, the content of professional training is framed by the state requirements for minimum content and the level of competencies for the university graduates in a particular area. The third level of higher education can be one of two types: the two year Masters program which prepares graduates for research and/or research–teaching, or the one-year (traditional) program, known as *specialitet*, that prepares students to be, for example teachers, engineers, economists.

Technology education is the program for students who want to graduate with a Bachelor of Technology Education (BTechEd) degree (four year program) or with a Master of Technology Education degree (six year program). State Standards (The Ministry of Education of the Russian Federation, 2000) specify the structure, length and content of the programs, the qualities of the graduates and requirements for the organisation of the education process at the university.

The technology education program has two profiles: technology of resistant materials and technology of textiles and food. The structure of the programs is discussed in detail in the next section. The Bachelor of Technology Education graduates have been prepared to teach in different types of education institutions. They can become involved in the following professional activities: teaching, research, socio-pedagogical, up-bringing, culture-enlightening, correction and development, and management (The Ministry of Education of the Russian Federation, 2000).

An important part of the Master of Technology Education program is its research base. The Standards specifies 1854 hours of research–related activities out of 3888 total hours. After graduation, these students can continue their study in *aspirantura* at the post-graduate level. Aspirantura is a traditional post-graduate institution that operates within universities and Academies of Sciences. It is a research-based study that ends up with a PhD thesis.

In 2003, the all-Russia classification of qualifications was approved with implementation from the 1st of January 2004. It introduced the qualifications in technology education outlined in Table 3.

Table 3.
Qualifications relevant to technology education.

Title	Qualification
Technology education	Bachelor of technology education
	Master of technology education (workshop teacher in VET)
Professional education (in different areas)	Teacher of vocational education
	Specialist of industry training
Technology and Enterprise	Teacher of technology and enterprise
Technology	Technology teacher

AN EXAMPLE: HERZEN STATE PEDAGOGICAL UNIVERSITY

The program described here is the current program at Herzen State Pedagogical University in St. Petersburg, which is considered one of the best among the pedagogical universities in Russia. There are several education programs at the faculty, including Bachelor and Master of Technology Education (full time) and Technology and Enterprise (distance study and short on-campus sessions) that prepare technology teachers. This faculty pioneered the multi-level technology teacher-training program that consists of the BTechEd, which is four years plus one year of *specialitet* (full-time). After its introduction in 2002, the program has subsequently extended throughout Russia.

While studying at the Bachelor level within the technology education program, students can choose one of two profiles: technology of resistant materials or technology of textile and food, to focus their study in accordance with the separate courses for boys and girls taught at the school level. Table 4 presents an example of the multi-level training.

Technology Teacher Education in Russia

Table 4. Bachelor of Technology Education program.

Name of the discipline (and practicums)	Hours			Semesters								Assessment
	Total	At the Uni	At home	1	2	3	4	5	6	7	8	
Humanities and socio-economic disciplines	**1510**	**954**	**546**									
Federal component	1050	954	546									
Foreign language	430	170	170	X	X	X	X					Exam
Physical culture	408	408		X	X	X	X	X	X			Pass
Russian history	100	50	50	X						E		xam
Philosophy	102	52	50		X	X						Exam
Economics	100	50	50			X						Exam
National-regional component (University)	225	112	113	X	X	X						Pass
Elective disciplines and courses of students' choice	225	112	113	X	X				X			Pass
Mathematics and science	**1000**	**515**	**485**									
Federal component	800	515	485									
Mathematics	220	110	110	X	X	X						Exam
Informatics	110	56	54	X	X							Pass
Physics	220	110	110	X	X	X						Exam
Chemistry	150	74	76	X	X							Pass
Ecology	100	57	43	X								Pass
National-regional component (University)	100	54	46	X	X							Pass
Elective disciplines and courses of students' choice	100	54	46			X	X					Pass
General professional disciplines	**2794**	**1458**	**1336**									
Federal component	1676	900	1776									
Psychology	400	202	198	X	X	X	X	X				exam
Pedagogy	300	150	150	X	X	X	X					exam
Basis of the theory of technology education	300	162	138						X	X	X	exam
Educational technology and methods of teaching (according to the two profiles below)	200	114	86						X	X	X	exam
Basis of research in technology education	100	56	4						X			Pass
Applied mechanics	300	179	130					X	X	X	X	Exam
Theory of machines	240	134	106					X	X	X	X	Exam
Safety through life	36	18	18	X								Pass
National-regional (university) component	559	279	280	X	X	X	X					Pass
Elective disciplines	559	279	280	X	X				X	X		Pass
Disciplines for profile Technology of resistant materials	**1600**	**903**	**697**									
Organisation of the modern industry	300	150	150						X	X	X	Exam
Technical disciplines (Disciplines of technological cycle)	400	250	150				X	X	X	X	X	Exam
Electro-radio-technology	300	182	118						X	X	X	Exam
Practicum on technology of resistant materials	300	150	150			X	X	X	X			Exam
Technical creativity	300	150	150			X	X	X	X			Pass
Disciplines for profile Technology of textile and food	**1600**	**929**	**671**									
Management and technology of the service providing industries	200	100	100						X		X	Pass
Technology of resistant materials	100	54	46						X	X		Pass
Electro-radio-technology	300	182	118						X	X		Pass
Practicum of sewing technology	300	144	156						X	X	X	Exam
Basis of micro-biology, physiology and hygiene of food	100	57	43			X	X					Pass
Practicum on culinary	300	200	100					X	X	X	X	Exam
Special drawing	100	72	28					X	X			Pass
Decorative applied art	200	120	80		X	X	X	X	X			Pass
Facultatives Military training	**450**	**225**	**225**					X	X	X	X	Pass
Total	7344	4055	3289									

Pavlova

After finishing their bachelor studies, students can study for a further year to become teachers of technology and enterprise education. There are two specialisations within that qualification: *Technic and Technical Creativity,* and *Culture of the House and Craft-applied Art Creativity.* The study program for technic and technical creativity is summarised in Table 5.

Table 5. Program for 1 year *specialitet.*

Name of the discipline	Exams	Pass	Projects	TOTAL	Total at Uni	Hours lectures	Hours tutorials	Hours labs.	Hours homework	Semesters 9 (10 weeks)	Semesters 10 (10 weeks)
	Semester	Semester									
HUMANITIES AND SOCIO-ECONOMIC DISCIPLINES				120	60	30		30	60		
Culturology		9		40	20	10		10	20	101	
Politology		9		40	20	10		10	20	101	
Sociology		9		40	20	10		10	20	101	
GENERAL MATHEMATICS AND SCIENCE DISCIPLINES				194	40	20		20	154		
Mathematics • Math statistics in psychology-pedagogy research		9		114	20	20		10	94	101	
Physics • Modern technical physics				80	20		10	10	60	101	
GENERAL PROFESSIONAL DISCIPLINES				256	100	50		50	156		
Basis of specialised pedagogy and psychology	10			72	40	20		20	32		202
Theory and methods of teaching technology and enterprise • Methods of teaching enterprise			10	40	20	10		10	20		101
Developmental anatomy, physiology and hygiene		9		72	20	10		10	52	101	
Basis of medical knowledge		9		72	20	10		10	52	101	
DISCIPLINES OF THE AREA				764	270	110		160	494		
Theory of machines	10			60	20	10		10	40	101	
Technology disciplines • Technology of modern materials • Automatic systems of management				120	50	20		30	70		102
Information technology		9		50	30	10		20	20	201	201
Graphics • Graphic design		10		100	30	20		10	70	102	202
Basics of entrepreneurial activities: • Marketing • Finance and finance analysis	9			280	80	40		40	200	202	202
Basis of creative engineering activities		9		40	20	10		10	20		101
Technology practicum • Practicum in the workshops; • Life safety in technological environment.				114	40			40	74	004	
Disciplines of specialisation				194	70	30		40	124		
Current research in theory of machines		10		54	20	10		10	34		101
Current home appliances		10		100	30	10		20	70		102
Robotics		10		40	20	10		10	20		101
Number of study hours				1528	540	240		300	988	27	27
Projects				1						1	-
Number of exams				4		2				-	2
Number of passes				15						8	7
Final state attestation for graduation											
• Graduation project – Project on professional activity of technology and enterprise teacher • State exam – Pedagogy and methods of technology education											
Students also have 9 weeks of pedagogical practice in semester 9 and 9 weeks of industry placement in semester 10.											

These new educational programs within the Faculty have been designed in accordance with the program for modernisation of Russian education – they are less oriented towards the transmission of knowledge, and more towards the development of the interests and creative capabilities of all students, towards stimulating their independent work/study, and developing experiences within creative activities. Pedagogy is student-oriented. However, these programs do not prepare students for teaching design in school, although a draft of the second Standards for school programs introduced design as an important part of the general component for all students to learn.

A 1999–2000 study (Korshunova & Korshunov, 2001) indicated that 63.5% of the graduates work in education, 53% as technology teachers in secondary schools and the remainder in management and the economy, including service industries, social spheres, other industries, and staying at home to look after children. This data highlights that around 66% of all graduates are not working as technology teachers, but found employment in other areas. This is possibly due to the very broad training program that goes back to the general, theoretic, systematic, structured tradition in education that prepares graduates to take up a range of opportunities. One of the main reasons for not taking up teaching jobs is the low status and salary of teaching.

In relation to the students' satisfaction with the university training program, the same study (Korshunova & Korshunov, 2001) revealed that 71% of the current 4th year students were happy that they had chosen this program. However, both the 1999 and 2000 cohort identified the need for a more balanced, all-round education.

Respondents highlighted their desire to have more practical knowledge and skills, thus to have more practicums in workshops, industry, and schools, more practical foreign language courses, and computer skills. It is also evident from the data that students want to develop their knowledge in humanities, believing that technical and mathematical disciplines are adequately covered in the program.

This attitude towards the technical disciplines (that had been taught in the program at the level of engineering faculties) was identified as separate from future professional activities, by almost half of the respondents. The majority of negative responses were from female graduates who had not applied this particular knowledge in their teaching (home economics context). Recent interviews with the dean of the faculty (March 2004) also identified restructuring of the engineering courses as the most significant current issue within the program.

These training programs provide a suitable basis for further professional development of graduates. According to Korshunova & Korshunov (2001), 9% of graduates studied at the post-graduate level and 26% were enrolled in professional development studies. Of the graduates, 14% have a second higher degree, while 30% graduated from a variety of other courses (including foreign language, management, psychology, accountancy, and building).

Thus, this research identified overall satisfaction with the program with some suggestions to increase the practical orientation of training and the humanities component, and to decrease the amount and depth of technical-engineering disciplines. However, this program produced graduates who can succeed in the competitive labour market due to the breadth and fundamental nature of their education.

TEACHER CERTIFICATION

When students graduate from the university course with qualifications such as Teacher of Technology and Enterprise, Technology Teacher and Teacher of Vocational Education, their degree automatically gives them the right to teach in schools. There is no extra certification or registration process. However, to graduate, students must pass the state graduation exam and defend their graduation project. This is done to identify the level of practical and theoretical preparation of each technology teacher for fulfilling the professional tasks specified in the State Standards.

It is worth mentioning that a well-established system of professional development for teachers operates in Russia. After graduation, each teacher must be periodically enrolled in a professional development program taught by special universities or institutes. The usual arrangement is one day of study per week during the year or several two-week sessions during the year every five years.

Another unique feature of technology teacher training in Russia is an established comprehensive network of Heads of Technology in the education departments within Universities (currently 57) all over Russia. They meet regularly to develop common policy and share effective practice in the area.

CONCLUSION

In this chapter the training of technology education teachers has been examined within the context of the government program for modernisation of the Russian education system. Several trends have emerged. Technology teacher training in Russia is very centralised. Programs are regulated by the Standards and developed at the federal level with the possibility of up to 5% variation from the regular specifications. Specialisations within these programs provide some flexibility in terms of training. There are a limited number of specialisation programs which are required to be approved by the Learning–Teaching Methodology Unions. The current major trend in developing training programs is the introduction of Bachelors and Masters Degrees within the traditional five-year training system.

All training programs continue to be very broad in nature, aimed at the development of a well-educated, all-round person, so graduates can work in any areas, not just as technology teachers. However, more balance is required: for example, the inclusion of more humanities studies and a decrease in the amount and depth of engineering courses. Programs remain very theoretical, so there is a need to include more practical courses and to challenge the paradigm that theory should always precede practice.

The major future direction for technology teacher training programs is seen by the author as developing along a path consistent with the ideology of modernization, which means an orientation towards problem-based learning. The argument for this approach is framed with the belief that the training program not only provides systematic understanding of the physical/technical world, but considers it in terms of technology's relationship with society and nature. In terms of pedagogy there is a definite shift towards more student-oriented learning. A movement from directed technical education towards a design-based education is seen as the way for further program development. This initiative is due to the introduction of design-based approaches at the secondary level.

REFERENCES

The Federal Component of the State Standards for General Education (draft)(2003), retrieved from the web 2004, http://www.ndce.ru/ndce/Min_Edu/stand/index.htm

Korshunova, N.N. and Koeshunov, T.Y. (2001). *Professional portrait of the graduate faculty of technology and enterprise.* St.Petersburg: University Publisher

Lednev, V. S., Nikandrov, N. D., & Lazutova, M. N. (Eds.). (1998). *Uchebnue standartu shkol Rossii. Gosudarstvennue standartu nachalnogo obstchego, osnovnogo obstchego I srednego (polnogo) obstchego obrazovanija. Kniga 2. Matematika I estestvenno-nauchnue distsiplinu* [Learning Standards for Russian Schools. State Standards for primary and secondary education. Book 2. Mathematics and Science]. Moscow: Sfera, Prometej.

The Ministry of Education of the Russian Federation (2003a). *The principal findings of the study into the implementation of the strategy of modernisation of Russian Education in 2002 – Analytical report.* Moscow: Ministry of Education of the Russian Federation

The Ministry of Education of the Russian Federation (2003b). *Order 4482 from 04.12.2003. About the use of All-Russia classification of specialities in education,* The Author: Moscow

The Ministry of Education of the Russian Federation (2000). *State Educational Standards for Higher Vocational Education. Speciality 030600 Technology and Enterprise, qualification: technology and enterprise teacher.* The Author: Moscow.

The Ministry of Education of the Russian Federation, National Fund for Personal Training (2001). *Strategija modernizatsii soderzhanija obschego obrazovanija* [Strategy of modernisation of general education]. Upravlenie shkoloj, 30, pp. 2–16.

Pavlova, M., Pitt, J., Gurevich, M., Sasova, I. (2003). *Project method in technology education of students.* Moscow: Ventana-Graff.

Piskunov, A.I. (2001). The history of pedagogical education: from the appearance of up-bringing in primitive societies to the end of the XX century. Sfera, Moscow.

The program of vocational education development, retrieved from the web 2004, http://www2.ed.gov.ru/prof-edu/sred/rub/

Russian Education by 2001: Analytical overview, retrieved from the web 2004, http://www2.ed.gov.ru/uprav/obzor/2.5

Russian Education: Federal Portal, retrieved from the web 2004, http://www.edu.ru/gw/db.informika.ru/cgibin/portal/3nk_retro/list_1_1.pl?spe=030600&fokr=&ter=&okato=&okogu=&fs

Review of vocational education development in 2001 –2002, retrieved from the web 2004, http://www2.ed.gov.ru/prof-edu/sred/rub/

Russian Government (1994) *Order N 940 from 12.08.1994. About the approval of the State Educational Standard for the Higher Vocational Education,* retrieved from the web 2004, http://www.edu.ru/db/portal/spe/docs/1994_940.htm

Statistics of Russian Education, retrieved from the web 2004, http://stat.edu.ru/stat/vis.shtml

UMOa, retrieved from the web 2004, http://mpgu.edu/umo/02/zadachi.html

UMOb, retrieved from the web 2004, http://mpgu.edu/umo/

Vocational education (2003) http://www2.ed.gov.ru/prof-edu/nach/rub/zp/246,print/

Technology Teacher Education in South Africa

Chapter
10

Chris Mothupi
PROTEC, South Africa
Andrew Stevens
Rhodes University, South Africa

INTRODUCTION

In 1994, after nearly five decades of oppressive apartheid rule, South Africa became a democracy and took its place amongst the free nations of the world. The new Constitution, a product of intense debate over a considerable period, established education as a basic right and for the first time school attendance became compulsory for the first nine years of schooling (ages 6–7 to 14–15). Later legislation reorganised education and training into three bands: a General Education and Training (GET) Band, encompassing the compulsory years (Grades 1–9), a Further Education and Training (FET) Band, comprising the final three years of secondary school (Grades 10–12) and a Higher Education and Training (HET) Band which includes university education and education in newly named 'universities of technology' (formerly known as 'technikons'). In line with the adoption of a new outcomes-based philosophy of education, completely new curricula were drafted to fit this new conception of education. Although the curricula reflect these new 'Bands', schools are still organised according to the traditional Primary (Grades 1–7) and Secondary (Grades 8–12) arrangement. Administratively, public education is the responsibility of nine provincial departments of education whose function it is to pay teachers, maintain and supply schools, provide curriculum support for teachers and manage the external assessment of learners at the two main exit points, Grade 9 and Grade 12.

In sum, the South African public schooling system comprises close to 12 million learners, taught by 340 000 teachers in 26 000 schools. The independent school sector (some of which receives a measure of government support) forms a small but significant sector, accounting for a further 280 000 learners in 1 100 schools with 16 000 teachers. (South Africa, Department of Education, 2004). Teacher education takes place in the higher education sector where the total student enrolment is close to 700 000 in 23 institutions. The country spends nearly 6% of its Gross Domestic Product (and over 20% of its annual budget) on education, one of the highest proportions in the world: yet, in the school sector, over 90% of this budget is spent on salaries and wages, leaving little for capital expenditure and other resources. The majority of South Africa's schools remain poorly resourced and difficult places in which to offer quality learning experiences.

Technology Education in South Africa is new in the school curriculum, being introduced in 1998. Technology is a compulsory learning area up to Grade 9, although it does not exist independently in the Foundation Phase (to Grade 3) in which it is part of the Life Skills Learning Programme together with Literacy and Numeracy. The Technology Curriculum consists of Learning Outcomes and associated Assessment Standards. Learning Outcomes specify the core concepts, content and skills for each grade level and Assessment Standards describe the minimum expected level of performance at each grade level.

Learning Outcome 1 in technology is the core outcome and lists the assessment standards for technological skills: *The learner is able to apply technological processes and skills ethically and responsibly using appropriate Information and Communication Technologies (ICT)*. The Assessment Standards in this outcome are organised under five technological skills: Investigating, Designing, Making, Evaluating and Communicating. Each of these technological skills has a number of specific assessment standards attached to it. All these Assessment Standards, along with those for Learning Outcomes 2 and Learning Outcome 3, could be achieved through project-based learning experiences that expose learners to all aspects of the learning area in an integrated way.

Learning Outcome 2 in technology lists the assessment standards for Technological Knowledge and Understanding: *The learner is able to understand and apply relevant technological knowledge ethically and responsibly*. The Assessment Standards in this outcome are organised under three content areas: Structures, Processing and Systems and Control (Mechanical and Electrical/Electronic Systems). Learners could demonstrate achievement in these outcomes in the process of completing practical project work.

Learning Outcome 3 lists the assessment standards for Technology in Society: *The learner is able to demonstrate an understanding of the inter-relationships between technology, society and the environment.* It is organised under the headings Indigenous Technology and Culture, Impacts of Technology and Bias. Learners could demonstrate achievement in these outcomes in contexts directly related to practical project work and in particular through investigating and evaluating particular aspects of project work.

Table 1.
Assessment Standards for Technology

Outcome 1	Outcome 2	Outcome 3
Investigating 1.1 Finds information re a problem or need. 1.2 Investigates (evaluates) existing related products. 1.3 Performs practical tests. 1.4 Uses appropriate information technology.	**Structures** 2.1 How materials can be used to make and reinforce structures.	**Indigenous Technology** 3.1 How different cultures have used materials, solved problems and optimized solutions.
Designing 1.5 Writes a clear statement summarizing the need. 1.6 Writes specifications 1.7 Suggests possible solutions. 1.8 Chooses one with reasons.	**Processing** 2.2 How materials can be processed to improve their properties.	**Impacts** 3.2 How solutions impact on society and the environment.
Making 1.9 Planning. 1.10 Making skill. 1.11 Safety and working practice. **Evaluating** 1.12 Evaluates products and suggests improvements. 1.13 Evaluates the process and suggests improvements. **Communicating** 1.14 Produce sketches and drawings. 1.15 Make presentations using ICT.	**Systems and Control** 2.3 How mechanical systems work to gain advantage. 2.4 How electrical/ electronic systems work	**Bias** 3.3 How solutions might disadvantage certain groups and possible ways of redressing bias.

HISTORY

Prior to 1994, education in South Africa was organised on racial lines with separate schools, universities, teacher colleges and administration systems for each of the four main groups as defined by the apartheid state: namely black, white, coloured and Indian. To complicate maters further, there were four so-called 'independent homelands' within the borders of the country for four of the main black population groups, each having their own educational ministry and administration. Although the curricula in each of these systems was theoretically equal, the huge differential in state funding made a mockery of the apartheid state's claim of 'separate but equal' treatment for all races.

It is perhaps not surprising that it was an education issue which sparked the famous 'Soweto Uprising' in June 1976 which signalled the beginning of the end of apartheid. The fifteen years following 1976 were marked by almost continual unrest in many of South Africa's black schools which was met with considerable force by the state. Military vehicles were a common sight on school property in this period. One of the more considered responses of the state was a comprehensive investigation into education published as 'Provision of Education in the RSA' by the Human Sciences Research Council (HSRC, 1981) which attempted to shift the focus on formal education based on the traditional 'academic' arts and sciences curriculum towards a more 'appropriate' skills-based vocational curriculum, particularly for the majority of black school goers (Kraak, 2002).

In the eighties, a period characterised by widespread unrest in black education, the vocationally oriented ideas of the HSRC report began to take root in much of the official discourse of the time. The Education Renewal Strategy (ERS) recommended the introduction of a number of new compulsory subjects into the general formative curriculum (South Africa. Department of National Education, 1991). Amongst these were the new (to South Africa at least) subjects: Economics, Technology and Arts Education, the rationale being that these three subjects would provide education relevant to the needs of learners and society as well as contributing to the person-power requirements of the country.

The ERS seems to have been aware of research pointing towards the dangers of vocationalising or specialising too early, and it was careful not to propose too clear a differentiation between academic and vocational pathways in the compulsory phase of secondary education (that is,

Grades 1 to 9), but, in the proposed post-compulsory phase, vocational education was to assume far greater significance. Although never implemented, these proposals by the apartheid government were to find strong echoes in the policies and legislation of the new democratic order.

When the ANC convincingly swept to power in the first democratic elections in 1994, much was expected, particularly in the long neglected area of educational transformation. At the level of policy, Kraak (2002) identified the following three pillars underpinning the new dispensation:

- An integrated education and training system. The new government committed itself to eradicating the difference in status and privilege which a differentiated 'academic' versus 'technical/vocational' system promoted.
- A single qualifications structure. A new statutory body, the South African Qualifications Authority (SAQA) was established in 1995 to co-ordinate and manage the new National Qualifications Framework (NQF).
- A new curriculum framework. The new curriculum, named Curriculum 2005 (C2005) for the year in which implementation was to be accomplished, is the first single curriculum for all South Africans. Education, for the first time, was to be compulsory for all learners for nine years, the newly named General Education and Training (GET) Band. Thereafter would follow three years of Further Education and Training (FET) which would provide for more differentiated general, vocational and work-based education and training.

A feature of the Curriculum 2005 has been the introduction of eight new compulsory 'learning areas' (replacing the label 'subjects' was an attempt at encouraging the integration of disparate 'disciplines'). These were Language, Literacy and Communication; Mathematics; Human and Social Sciences; Natural Sciences; Technology; Arts and Culture; Economics and Management Sciences; and Life Orientation. For the first time, Technology, in a form corresponding largely to the British Design and Technology model, was to be part of every learner's education to Grade 9.

This curriculum represented a radical break with the past: in addition to embodying the broad political aims of access, equality and redress, the new curriculum was designed around the 'new' philosophy of constructivism and 'outcomes-based education' (OBE). Among the many new features to the South African school curriculum education were the notions of integration of discrete 'subjects' into broader 'learning areas', the idea of learner-centredness and the teacher as facilitator (rather than sole authority), and the policy of continuous assessment rather than the reliance on final examinations to determine progression through the system. One of the eight new learning areas to be introduced was Technology, in spite of the fact that very few teachers had received any formal training to teach this learning area.

Underlying the whole educational system are twelve 'critical outcomes' which all learning programmes are presumed to encompass. These include problem solving, working co-operatively, time management, communication in various modes, and using science and technology effectively. The new Technology learning area, itself a product of recent educational thinking, clearly incorporated most of these critical outcomes, and was certainly a factor which prevented it from being removed from the curriculum when the curriculum was reviewed in the year 2000.

Another important feature of the new system with a bearing on technology education is the introduction, after the formative general education of the GET Band, of specialised education at the FET levels. The new subjects introduced at this level include Design, Computer Applications Technology, Mechanical Technology, Electrical Technology, Civil Technology, Engineering Graphics and Design, and Consumer Studies.

After three years of implementation of the new Curriculum 2005 and the new approach to teaching (OBE), it became clear that teachers were not coping with those changes, partly because the training that they received was inadequate in general, and for Technology, it was overly theoretical and focused more on understanding the curriculum policy. The reaction of the ministry of education to the situation was to appoint a Curriculum Review Committee to evaluate the implementation of C2005. One recommendation of the committee was to scrap Technology completely, but after a concerted national and international response, this recommendation was not implemented.

OVERVIEW OF TECHNOLOGY TEACHER EDUCATION

Teacher education in South Africa, like the whole of the education system, is in transition. After the African National Congress (ANC) won the country's first ever democratic elections in 1994, the new government focused on the need to develop a post-apartheid school curriculum. After a range of enabling legislation had been put in place (South African Qualifications Act of 1995, National Education Policy Act of 1996, South African Schools Act of 1996, amongst others), the new Curriculum 2005, was introduced at the beginning of 1998. For the first time school attendance was to be compulsory for all South African children for the newly named General Education and Training (GET) Band, which covers the first nine years of schooling.

Once the new GET school curriculum had been launched, attention could be turned to other areas of the system such as the Further Education and Training Band (Grades 10–12) and Higher Education and Training under which teacher education resided. In teacher education particularly, the new government inherited a complex and chaotic situation. Under apartheid, teacher education took place largely in colleges of education which were conceived of as institutions located somewhere between secondary and tertiary institutions and were administered by a range of national and provincial departments, depending on the race of the group for which they catered. Although there was tight ideological control over these institutions, there was little or no emphasis on quality assurance or even on the numbers of teachers trained.

A national audit conducted in 1995 revealed that there were some 150 publicly funded institutions providing teacher education to over 200 000 students, including 70 000 initial teacher trainees at the 93 'contact' colleges alone. Considering that the South African system employs around 355 000 teachers, it is clear that too many teachers were being trained during the eighties and early nineties. A major restructuring of the sector has reduced the number of contact colleges to 25, and the number of distance colleges to two. All these institutions have been incorporated into universities and are now seen as an integral part of the higher education sector. This process, only recently completed, has coincided with recent legislation requiring all new teachers to be in possession of a degree before being employable in the state system. Prior to this a three-year college diploma had been sufficient to qualify as a professionally trained teacher and this remains the most common qualification of South African teachers.

When technology was introduced into schools as part of Curriculum 2005 in 1998, there were very few formally trained technology teachers in the system. Training had previously been received through either a nationally appointed task team or the few non-governmental organisations (NGOs) that had been involved in pioneering the introduction of technology into South Africa before it became part of the new curriculum. Although a few NGOs have managed to survive in the difficult funding environment of the 'new' South Africa, most of the formal technology teacher education takes place under the auspices of the higher education institutions. Here the major emphasis since 1994 has been on upgrading the qualifications of teachers who had been disadvantaged under apartheid education and providing in-service programmes in learning areas for which there are shortages, notably technology, mathematics and science. Such teachers may enrol for the two-year, part-time Advanced Certificate in Education (Technology Education) or, more recently, a three-year part-time Bachelor of Education (BEd) degree course. Most higher education institutions also offer the BEd degree as a four-year, full-time initial teacher qualification. Within these programmes, students will have the option to study Technology as one of their teaching method subjects.

Another option for students intending to become secondary or further education teachers is to enrol in a three-year degree programme (such as a B.A., B.Sc., B.Com.) and follow this with a 'capping' post-graduate certificate in education (PGCE). During this single-year, full-time course, Technology may be offered as a teaching method, but since very few South African universities offer suitable undergraduate courses in 'technology' or some suitable cognate subject, this is an option which has produced few technology teachers.

Although technology is now firmly established in the compulsory GET phases, the recent restructuring of the FET (Grades 10–12) sector has resulted in the abandonment of the generalist and integrative approach of the GET Band and given the last three years of schooling a markedly specialised and vocational slant. From 2006, learners interested in technology-orientated fields will, at the relatively early age of 15–16 years, be required to choose to pursue studies in one or more of the following five subjects: Electrical Technology, Mechanical Technology, Civil Technology, Engineering Graphics and Design and Computer Applications Technology. Those who prefer a 'softer' approach to technology may choose the new subject 'Design' which has emerged from the Arts and Culture traditions and most closely resembles the technology learning area of the GET Band. The five specialised technology subjects appear to be reformulations of the technical subjects which were offered in the vocationally-orientated technical schools and colleges of the past: it is doubtful whether many ordinary secondary schools, which form the vast majority of schools in the country, will have the capacity, either in terms of physical or human resources, to be able to offer these subjects as part of their programmes. The teacher education challenges posed by these changes have yet to be faced since many of the colleges which provided this specialised training no longer exist or have been incorporated into higher education institutions which do not have a tradition of skills-based education and training.

As a result of the lack of pre-service training of teachers, most of the current technology teachers have been trained though in-service programs provided by provincial education departments, some institutions of higher education and non-governmental organisations. As a way of addressing the challenges of having a subject in the curriculum and not enough educators to teach it, the National Department of Education put out tenders for service providers to train educators in order to enable them to teach Technology in schools, the result of which produced a hybrid of technology teacher training providers, made up of NGO's and universities. At present the whole field of teacher education has been the subject of a two-year study by a specially appointed Ministerial Committee on Teacher Education whose findings have recently (June 2005) been published.

STRUCTURE OF TECHNOLOGY TEACHER EDUCATION

The period after the first democratic elections in 1994 saw a dramatic restructuring of teacher education in South Africa. The first five years of democracy was a period of intense policy formulation and saw a large volume of legislation being enacted. The first of these was the South African Qualifications Authority Act of 1995 which created for the first time a National Qualifications Framework (NQF) and an authority (SAQA), independent of the Department of Education, to register qualifications and administer the framework. By this act, space was created for private providers of education and training to offer a wide variety of courses and qualifications (including teacher education courses) previously the sole preserve of the state-funded colleges and universities. This was followed by further legislation which placed teacher education firmly in the sphere of the Higher Education sector, thus removing it from the control of the provinces and placing it under the national Department of Education. The

National Education Policy Act of 1996 gave the national Minister of Education the power, amongst other things, to determine policy for:
- the professional education and accreditation of teachers;
- curriculum framework, core syllabuses and education programmes, learning standards, examinations and the certification of qualifications, subject to the provisions of any law establishing a national qualifications framework or a certifying or accrediting body.

As Parker (2003) notes, the Act gives the Minister clear authority over teacher education, but the responsibility for policy on teacher education programmes and qualifications is diluted by the reference to the NQF which is under the authority of SAQA. Such overlapping functions and responsibilities are perhaps inevitable given the newness of South Africa's democracy, but these and other confusions have not aided the pace of policy implementation.

Perhaps more significant for the movement to transform teacher education was the publication as national policy of revised *Norms and Standards for Educators (NSE)* and *Criteria for the Recognition and Evaluation of Qualifications for Employment in Education (CREQ)* in 2000, which provided the basis on which qualifications and learning programmes for educators could be developed. The publication also spelled out in some detail the types of qualifications that the Department would consider for employment and subsidy funding. These new policy documents placed teacher education firmly within an outcomes-based epistemological framework and represent a radically different way of conceiving and structuring teacher education. For the first time in South Africa's history, a framework and a procedure for approving and evaluating teacher education programmes were created. While not prescriptive as to content, a cornerstone of the policy is seven roles and their associated applied competences that have to be built into all teacher education qualifications. The seven roles that teacher programmes are meant to develop are those of:
- learning mediators;
- interpreters and designers of learning programmes and materials;
- leaders, administrators and managers;
- scholars, researchers and lifelong learners;
- community, citizenship and pastoral role players;
- assessors;
- learning area/subject/phase specialists.

Within each of these roles students are to develop 'applied competence', an overarching term which is made up of three kinds of competence, namely practical, foundational and reflexive competence, which are defined as follows:

- *Practical competence* is the demonstrated ability, in an authentic context, to consider a range of possibilities for action, to make considered decisions about which possibility to follow, and to perform the chosen action.
- It is grounded in *foundational competence* where the learner demonstrates an understanding of the knowledge and thinking that underpins the action taken; and
- integrated through *reflexive competence* in which the learner demonstrates ability to integrate or connect performances and decision-making with understanding and with an ability to adapt to change and unforeseen circumstances, and to explain the reasons behind these adaptations (South Africa, Department of Education, 2000a, p.10).

From the language used in the descriptions above, it is clear that the Department intends all teacher education offered by higher education institutions to embrace the outcomes-based philosophy which underpins the new curriculum. To this end, the document develops in considerable detail how the various competences are to be manifested in the seven roles. A concern of the more traditional 'discipline-centred' educationalists, that the content knowledge of teachers was being underemphasised is addressed in the following extract:

> The seventh role, that of a learning area/subject/discipline/phase specialist, is the over-arching role into which the other roles are integrated, and in which competence is ultimately assessed. Qualifications must be designed around the specialist role as this encapsulates the 'purpose' of the qualification and 'shapes' the way the other six roles and their applied competences are integrated into the qualification. (South Africa, Department of Education, 2000a, p.12).

By the publication of these policies, the Department indicated its expectation that higher education institutions would redesign their courses and qualifications to reflect the new policies. The policy further specified what names would be used for various qualifications, what levels they would occupy on the NQF and what credit values they would carry.

Table 2 summarises the situation at the time of writing (2005), although there are recent indications that this is about to be further revised.

Table 2.
Higher Education Courses

Name	Aim	Credit	NQF Level
Certificate in Education	To develop introductory practical and foundational competence, and some degree of reflexive competence. To provide an entry or exit point before the completion of the Diploma in Education.	120	5
Diploma in Education	To accredit a learner with introductory practical, foundational and reflexive competence. To provide an entry and exit point before the completion of the BEd degree.	240	5
First Bachelors Degrees	To accredit a general formative qualification with one or more subject/learning area specialisations in order to provide access to a PGCE as a 'capping' qualification.	360/480	6
Post-Graduate Certificate in Education	To accredit a generalist educator's qualification that 'caps' an undergraduate qualification. As an access requirement candidates are required to have appropriate prior learning which leads to general foundational and reflexive competence. The qualification focuses mainly on developing practical competence reflexively grounded in educational theory.	120	6
Bachelor of Education	To accredit an initial qualification for educators in schools. The learner will have strong practical and foundational competence with the reflexive competence to make judgements in a wide context.	480	6
Advanced Certificate in Education	To accredit further specialised subject/learning area/discipline/phase competence, or a new subject specialisation, or a specialisation in one or more of the roles as an advanced study intended to 'cap' an initial or general teaching qualification.	120	6
BEd (Hons)	To accredit the advanced and specialised academic, professional or occupational study of an aspect of education. It is designed to build the competence of expert educators and curriculum specialists, system managers, or educational researchers.	120	7
Post Graduate Diploma in Education	To accredit advanced and specialised occupational, academic and professional study. This qualification can accredit the coursework component of a Masters' degree or provide and entry or exit point before the completion of a Masters' degree.	120	8
Master of Education	To accredit the advanced and specialised academic or professional study of an aspect of education with emphasis on research.	240	8
Doctor of Education	To accredit the highly advanced and specialised academic or professional study of an aspect of education in which the learner demonstrates capacity for sustained original research.	360	8

The following Table 3 locates the various teacher education qualifications on a hierarchical framework and provides a diagrammatic view of possible career paths for teachers and other students in Education.

Table 3.
Framework of Teacher Qualifications

NQF Level	Qualification and credit values
8	Doctor of Education (PhD) (360) Master of Education (MEd) By full thesis or course work and thesis (240) Post Graduate Diploma in Education (an exit point form the MEd) (120)
7	Bachelor of Education (Honours) 120
6	PGCE 120 ACE 120 BEd 480 First Degrees 360 or 480
5	Diploma in Education (240) Certificate in education (120)
4	FET Certificates (Grade 12 level)
3	School leaving certificates (Grade 11 level)
2	School leaving certificates (Grade 10 level)
1	GET Corticated (Grade 9 level)

Tables 2 and 3 give some idea of the structure of teacher education in South Africa at present, although recent work by the Ministerial Committee on Teacher Education (MCTE) suggests that further changes to the structure and functioning of the system are likely in the near future as the country faces a crisis in supply of teachers, particularly in the scarce Mathematics, Science, and Technology learning areas. Because of its recent introduction into the curriculum, Technology remains particularly vulnerable: its growth and development being stunted by a shortage of qualified teachers and by a shortage of teacher educators to provide the necessary initial and in-service education. The shortages of skilled personnel are evident also in the provincial departments of education where many posts in the advisory and specialist services in Technology remain unfilled.

As has been suggested above, the emphasis in teacher education since the introduction of the new curriculum has been on in-service teacher education. Here the most common qualification has been the Advanced Certificate in Education (Technology Education), ACE (TE). Typically, this 120 credit course will be offered over two years of part-time study, during which participants will attend lectures and workshops during holiday periods at higher education institutions. The actual structure and content of the course will vary from institution to institution, since higher education institutions are relatively autonomous, but recently introduced quality assurance regulation is encouraging institutions to share ideas and curricula approaches and may have the effect of bringing a semblance of order to what has been a somewhat disordered field. The ACE is also given in distance education mode with creative ways of accommodating the necessary practical components of the coursework. Although distance education is vital in South Africa's context, many remain sceptical about its efficacy in delivering quality educational outcomes in technology education, especially given the lack of resources available to the majority of the participants on such courses either at their homes or schools.

Another increasingly common option for in-service teachers is the newly introduced BEd (In-service) which is a generic qualification resulting in the award of a degree. This is available to teachers with a three-year diploma and will require them to undertake a further three years of part-time study. Unlike the ACE, this is not a specialist Technology Education qualification, but courses in Technology may be selected by participants as part of the qualification.

Although the focus in teacher education has been on in-service education in the decade since democracy, a looming shortage of teachers has shifted attention recently towards initial teacher education. As has been mentioned above, the education ministry drastically reduced the number of institutions offering initial teacher education (from over a hundred to twenty two) and incorporated the remaining colleges of education into higher education institutions. Although many of these higher education institutions offer courses in technology education, not all offer initial technology teacher education courses. Part of the reason is that subsidy funding has not encouraged higher education institutions to invest in the necessary infrastructure (equipment, workshops, laboratories) needed for initial teacher education (by its nature much more expensive than in-service education) and partly due to staffing shortages. Very few higher education institutions have been employing additional lecturing staff in the volatile educational environment of the last decade, preferring to use part-time staff to 'plug the gaps' until the situation stabilises.

In sum, technology teacher education is offered in some form at most of the twenty-three institutions in South Africa, although the emphasis is heavily in favour of in-service teacher education. In addition to the formal offerings of these higher education institutions valuable work is provided by non governmental organisations (NGOs) and other service providers which mainly offer short courses and workshops, although some of these providers do team up with higher education institutions to offer accredited qualifications. Initial teacher education is currently receiving necessary attention, although higher education institutions will need to increase their staffing and resource capacity to make an impact on the looming teacher shortages (some estimates place the shortage as high as 15 000 per annum) which are predicted.

AN EXAMPLE: RHODES UNIVERSITY

Rhodes University was one of the first South African universities to offer a formal qualification in Technology Education. In collaboration with the ORT-STEP Institute, an NGO which was a pioneer of technology in South African schools, the university developed a course for in-service education of GET Band (Grades 1–9) teachers. The course, originally called the Further Diploma in Education (Technology) but now termed the Advanced Certificate in Education (ACE) was initially taught by a combination of ORT-STEP and university staff members. It is a two-year part-time course which is only available to qualified teachers who have a minimum three-year College of Education diploma. The course is an example of the productive partnerships (between the formal and non formal sectors) which were a feature of the early years of the transformation of the South African education system. In this case the university provided the accreditation and other curriculum resources while the NGO provided the technology education expertise and the funding for the establishment of a small but adequate workshop and computer laboratory.

Most importantly, the university has provided sustainability for the technology education unit, for while the NGO has ceased to exist, the university has established a post in technology education along with the more established mathematics, science and other learning areas.

Because technology is so new to the school curriculum, the ACE course is designed with novice teachers of technology in mind. Although some of the teachers who enrol are former technical subjects teachers (including woodwork and home economics) who have some knowledge and skill in these areas, the vast majority have no background in technology. Basic drawing and designing skills are particularly weakly developed, and few students have experience of tools and materials beyond the most rudimentary. Approximately half the course time is devoted to technology education issues and topics, while the other half is shared amongst basic mathematics, science and general education topics. The mathematics and science components are specifically intended to support the technology component rather than produce 'qualified' math or science teachers. The general education component addresses contemporary issues such as learning theories, outcomes-based education (and the philosophy underpinning the current curriculum), multi-cultural education, assessment of learning and other topics of relevance to teachers. The course includes 400 hours of contact time with a further 800 hours to be spent in school-and home-based tasks and assignments. The contact sessions take place during school holidays and the occasional long weekend.

A departure point for the course is a belief that technology occupies a unique position in the curriculum as it encompasses thinking, doing and valuing activities. The teaching methodology employed on the course seeks to reflect these dimensions and the lecturer models the various methodological approaches which can be used. Technology is intrinsically learner-centred and activity-based and a deliberate attempt is made to embody these principles and to avoid the traditional 'talk and chalk'. The following list outlines the range of modalities employed:

- Small group work followed by plenary feedback and discussion.
- Co-operative group work where new concepts or terms need to be mastered/explored.

- Student-led peer teaching in order to practise new methodology.
- Lecturing with questioning and discussion.
- Practical sessions in which skills are learned and designs realised.
- Computer laboratory work where ICT skills are developed and research conducted.

As far as student assessment is concerned, a 'continuous assessment' policy is followed. It is stressed from the outset that all tasks need to be completed and assessed, whether formally (where the result is recorded and counts towards a year mark) or informally, where no mark is recorded, but the task's completion is noted. Tasks in the course generally fall into one of the following categories:

- case studies, where information on a particular product or context is gathered and presented.
- assembly/disassembly investigations.
- resource tasks, in which the primary purpose is to develop a particular skill or practical technique.
- capability tasks/assignments or projects, where students are expected to exhibit a range of skills across the researching, designing, making and evaluating phases of the technological process.

In practice, it is feasible to give no more than three fairly substantive projects per year. The results of these and a number of lesser tasks are used to compile a year mark which constitutes 50% of the final mark. A three-hour examination contributes the other 50%.

The course outcomes and content are in harmony with the school curriculum and reflect the themes and content areas of technological curricula globally (or at least those derived from the United Kingdom model). There is a strong emphasis on the design process, the acquisition of graphical skills, knowledge of a variety of materials, structures, systems and control, food technology and ICT. However, the curriculum also places emphasis on issues of sustainability and environmental impacts of technology as well as on learning about indigenous technologies and knowledge systems. A major tension in the course is the lack of emphasis on pedagogic content knowledge specific to the unique South African situation. This reflects the stage of development of the technology education field and the shortage of researchers working on such questions in South Africa at present. The emphasis and time of the relatively few academics involved in technology education has been on teaching, and the research output has been small.

Course evaluations by students have been positive. A frequent comment by students has been that they have found the course to be 'empowering' and that they have particularly enjoyed the practical workshop-based sessions. There is a sense that they have found 'relevance' in the course which was lacking in previous educational experiences. Whether they are able to pass on these positive experiences to their learners and so effect the necessary break in the cycle of educational 'disadvantage' remains to be seen. Much depends on the capacity of the provincial support systems whose efforts thus far have been less than ideal.

AN EXAMPLE: UNIVERSITY OF LIMPOPO AND PROTEC

An example of innovative Technology Teacher Training programmes are those run by the University of Limpopo and PROTEC in 2002 – 2005 called the 'Bushbuckridge Master Programme for Technology Teachers' and the 'Sekhukhuni Master Science and Technology Education Programme'. These were projects initiated by the National Department of Education in a response to the lack of properly trained Maths, Science and Technology (MST) educators in the Limpopo province. As a result of its MST strategy, the department put out tenders for the training of MST educators in some provinces. The University of Limpopo won the tender and contracted with PROTEC to develop and deliver the programme that would "empower technology educators with both content and pedagogic skills to boost educator confidence and their ability to motivate learners to participate and perform successfully in the Technology Learning Area" (DoE tender ED0212: 2001).

The objectives of the project were to empower technology educators with both content and pedagogic skills and to boost educators' confidence and their ability to motivate learners to participate and perform successfully in the Technology Learning Area. The curriculum focused on the

Senior Phase of the GET Band (Grade 7 to 9 teachers): the development of technological knowledge, concepts, associated technological skills and the acquisition of necessary pedagogical skills.

The course was split into 10 Modules with each module presented over a five-day period, that is, approximately 35 contact hours per module. Participants were required to complete classroom-based assignments that count towards accreditation. The modules were all self-contained and dealt directly with relevant content and methodologies for use in technology classrooms. These fall under the component titles Technological Studies (TS), Policy Studies (PS) and Curriculum Studies (TS).

Participants were exposed to the design process and were expected to design and make products of technology in the technology studies component. The design process was used as the vehicle for delivering key content, reinforcing OBE principles, studying learning theories, assessment strategies and as a springboard for planning activities falling under the Policy Studies and Curriculum Studies components.

Participants were expected to prepare, plan, present, assess and evaluate classroom based teaching projects during this course. These projects formed the basis for the final accreditation of the educator. By the end of the course, teachers demonstrated the ability to:

1. Produce a detailed Technology learning programme for the Senior Phase.
2. Develop technology learning experiences for Senior Phase learners.
3. Deliver and assess technology learning experiences for Senior Phase learners.
4. Evaluate the effectiveness of a learning experience and suggest adaptations.
5. Demonstrate the ability to use the following technological skills: investigate, design, make, evaluate, communicate.
6. Demonstrate knowledge and understanding of the key topic areas of Systems and Control (electrical/electronic, mechanical and pneumatic/hydraulic systems), Structures and Processing.

7. Demonstrate awareness and sensitivity to issues relating to technology and society: indigenous technology and culture, impacts of technology, technology and bias.
8. Develop and use assessment instruments to reliably measure and record the achievements of individual learners.
9. Demonstrate an ability to use resources of technology effectively.

Lectures were held during school holidays for a period of one or two weeks. Each contact session had a set of outcomes to be achieved by all students. Students were taken through activities contained in the module for each theme.

Assignments from previous workshops were marked and returned to students at the next workshop and became the basis of evaluative discussions. During the sessions, group discussion and brainstorming was utilized, based on for example, identified problems, possible approaches to solve problems, and the roles to be played by each group member. The 'making' phase of the projects incorporated group portfolio development, and concluded with testing and evaluation in terms of the brief and specifications, and then presentations to the rest of the class.

TEACHER CERTIFICATION

In order to assure the quality of its teacher education qualifications, South Africa has established a number of statutory and regulatory bodies. All providers of teacher education are required to register with and gain the approval of the Minister of Education, their programmes and/or qualifications need to be registered with the South African Qualifications Authority (SAQA) on the National Qualifications Framework (NQF) and they have to be accredited by the Higher Education Quality Committee (HEQF), which is a standing committee of the Committee on Higher Education. Once teachers have graduated with a recognised and accredited qualification, they are required to register with the South African Council of Educators (SACE) before they are given permanent appointment in a teaching post. This statutory body is responsible for the registration, promotion and professional development of educators and aims to enhance the status of the teaching profession. It is responsible for ensuring that its code of conduct is adhered to: teachers who contravene the code may lose their right to teach. SACE is also a key partner with the Department of Education and the Education Labour Relations Council in facilitating the upgrading of teacher qualifications.

CONCLUSION

Although seven years have passed since the introduction of technology into the South African curriculum, its place in the curriculum remains insecure. This has partly to do with a persisting ignorance of its nature and value, even amongst some educational officials and experts, many of whom still equate Technology Education with ICT. It is also partly to do with the massive challenges experienced in mathematics and science education, where the levels of performance of South African learners are amongst the lowest in the world. With limited resources to spend, most of these are used in attempting to address this significant problem.

Nevertheless, technology seems established in the GET curriculum at least, in spite of the fact that there is still a large shortage of qualified teachers and an even bigger lack of technological equipment and resources. There are positive signs that efficiency measures within the departments are freeing up funds, but it is unlikely that all schools will be provided with more than the very basic necessities in the short term. Computers are likely to consume most of the available budget.

A source of concern to many is the absence of technology (as a general subject) in the list of FET subjects, in spite of attempts by members of the Technology Association to petition the minister to include such a subject. It is the opinion of many that the four 'technology' subjects which have been included (Civil, Electrical, Mechanical Technology and Engineering Graphics and Design) are thinly disguised reformulations of the 'technical' subjects of the old curriculum and will require specialist teachers and workshops to teach them adequately. Few schools have the capacity to offer these specialist engineering-oriented courses, much as they may be needed. A general technology subject, it is felt, would be easier to resource and would provide the introduction to the design and technological skills needed by learners entering a job-scarce economy. This problem is exacerbated by the lack of easily identified higher education courses in 'technology' which prospective teachers might select as a foundation for teaching technology at secondary level.

At this stage in the development of the subject, it is important for newly qualified teachers to receive ongoing professional support. There is evidence that this is uneven in South Africa at present, with some provinces underperforming in this essential service. With little official support, the role of professional associations is vital. In this context, there is evidence of a vibrant culture of peer support emerging with the formation of at least two associations for technology teachers, both of which have regional branches and hold annual conferences.

Finally, the development of the field of technology education will depend on the combined efforts of teachers, teacher educators and researchers. With technology the newest of South Africa's new curriculum offerings, the field is open and inviting.

REFERENCES

Adler, J. and Reed, Y. (Eds). (2002). *Challenges of teacher development.* Pretoria: Van Schaik.

Chisholm, L. (Ed) (2004). *Changing class. Education and social change in post-apartheid South Africa.* Cape Town: HSRC Press.

Harber, C. (2001). *State of transition. post-apartheid educational reform in South Africa.* Oxford: Symposium.

Hartshorne, K. (1992). *Crisis and challenge: Black education 1910–1990.* Cape Town: Oxford University Press.

Human Sciences Research Council (1981). *Provision of education in the RSA.* Report of the main committee of the investigation into education. Pretoria: HSRC.

Jansen, J. & Christie, P. (Eds). (1999). *Changing curriculum. Studies on outcomes-based education in South Africa.* Cape Town: Juta.

Jansen, J. & Sayed, Y. (Eds). (2001). *Implementing education policies: the South African experience.* Cape Town: UCT Press.

Kahn, M.J. & Volmink, J.D. (1997). *A position paper on technology education in South Africa.* Johannesburg: Development Bank of South Africa.

Kraak, A. & Young, M.(Eds). (2001). *Education in retrospect: Policy and implementation since 1990.* Pretoria: HSRC.

Kraak, A. (2002). Discursive shifts and structural continuities in South African vocational education and training: 1981–1999. In P. Kallaway (Ed), *The history of education under apartheid, 1948–1994.* New York: Peter Lang.

Lewin, K., Samuel, M. & Sayed, Y.(2002). *Changing patterns of teacher education in South Africa: Policy, practice and prospects.* Sandown: Heineman.

Morrow, M. & King, K. (Eds). (1998). *Vision and reality: Changing education and training in South Africa.* Cape Town: University of Cape Town Press.

Mouton, J., Tapp, J., Luthuli, D.& Rogan, J. (1999). *Technology 2005: A National Implementation Evaluation Study.* Stellenbosch: CENIS.

National Education Policy Investigation. (1992). *Teacher education.* Report of the NEPI Teacher Education Research Group. Cape Town: Oxford/NECC.

REFERENCES

Parker, B.(2002). *Roles and responsibilities, Institutional landscapes and curriculum mindscapes:a partial view of teacher education policy in South Africa: 1990–2000.* In Lewin, K., Samuel, M. & Sayed, Y.(Eds). (2002). *Changing patterns of teacher education in South Africa: Policy, practice and prospects.* Sandown: Heinemann.

Robinson, M. (2003). Teacher education policy in South Africa: The voice of the teacher educators. *Journal of Education for Teaching*, 29(1) 19–34.

Sayed, Y. (2004). *The case of teacher education in post-apartheid South Africa: politics and priorities.* In Chisholm, L.(ed.). *Changing class: Education and social change in post-apartheid South Africa.* Cape Town: HSRC Press.

South Africa. Department of Education and Culture (DEC) (1990). *The Evaluation and Promotion of Career Education in South Africa.* Main Report of the committee chaired by Dr S.W.Walters. Pretoria: Government Printer.

South Africa. Department of National Education (DNE) (1991). *A curriculum model for education in South Africa.* Pretoria: Committee of Heads of Education Departments.

South Africa. Department of Education. (1995). South African Qualifications Authority Act No.58 of 1995. Pretoria: Government Printer.

South Africa. Department of Education. (1996). *National Education Policy Act No.27of 1996.* Pretoria: Government Printer.

South Africa. Department of Education. (1996). South African Schools Act No 84 of 1996. Pretoria: Government Printer.

South Africa. Department of Education. (1997). *Higher Education Act No.101 of 1997.* Government Gazette 18515. Pretoria: Government Printer

South Africa. Department of Education. (2000a). *Norms and Standards for Educators.* Government Gazette, Vol 415, No 20844. Pretoria: Government Printer.

REFERENCES

South Africa. Department of Education. (2000b). *Criteria for the Recognition and Evaluation for Employment in Education based on the Norms and Standards for Educators, 2000.* Government Gazette, Vol 423, No 21565. Pretoria: Government Printer.

South Africa. Department of Education. (2002). *Revised National Curriculum Statement Grades R–9 (Schools).* Pretoria: Department of Education.

South Africa. Department of Education. (2004). *Education statistics in South Africa at a glance in 2002.* Pretoria: Department of Education.

Stevens, A. (2004).Getting technology into the FET. In Buffler, A. & Laugksch, R. (Eds). Proceedings of the 12th annual meeting of the Southern African Association for Research in Mathematics, Science and Technology Education. Cape Town: University of Cape Town.

Stevens, A. (2005). Technology teacher education in South Africa. In *Technology Education and Research: Twenty Years in Retrospect.* Proceedings of the PATT-15 Conference. Available online at http://www.iteaconnect.org/PATT15/PATT15.html.

Technology Teacher Education in the United Kingdom

Chapter
11

Frank Banks
Open University, UK

INTRODUCTION

The Thatcher government of the late 1980s introduced a new and prescribed curriculum into state schools in England, Wales and Northern Ireland. For the first time anywhere in the world, Technology became a required subject for all pupils during the years of compulsory schooling from the ages of five to 16 years (See Smithers and Robinson 1992 for a first evaluation of this innovation). However, the nature of the school subject of Technology was not the same in terms of content and ethos in the four nations that make up the United Kingdom, and the precise level of prescription and the degree of compulsion for all pupils was also very different. Over the years, this national variation in what is embraced by the school subject area of 'Technology' has become more marked. Consequently, teacher education and training has a different emphasis in the different nation states.

The United Kingdom (UK) includes the countries of England, Northern Ireland, Scotland and Wales, along with a number of other minor states within the British Isles, such as the Channel Islands and the Isle of Man. Since 1999, responsibility for education has been increasingly devolved to these separate countries as national assemblies and, in the case of Scotland, a new parliament independent of London was created. In practice, both Northern Ireland and Scotland have been autonomous in the design and structure of their education systems for many years, and Wales is increasingly moving away from the curriculum model prevalent in England to allow the distinctive culture of Wales to be emphasized in the *Curriculum Cymreig*. Consequently, the task of describing the subject of Technology that is taught in UK schools is rather complicated as it has different formal curriculum requirements, a different status as a separate subject in its own right, and a different curriculum heritage in each constituent nation.

England, Wales and Northern Ireland have only had a state-specified curriculum, backed by the force of law, since 1988. In Scotland, the detail of the school curriculum is still not a legal requirement, but rather a series of 'guidelines' of what should be covered during the compulsory years of schooling. So, a brief outline of the nature of school technology in each country, in turn Northern Ireland, England and Wales, and Scotland, will be given.

In primary schools in Northern Ireland, Technology is formally combined with science as the subject of Science and Technology, although *making* remains an important element of that curriculum. An 'applied science' view of the Technology curriculum has been adopted in secondary schools where the subject is known as Technology and Design (T&D). It is available throughout post-primary school but is compulsory for lower school pupils (12 to 14 years old, known as Key Stage 3).

The central focus of Technology and Design at Key Stage 3 is the design and manufacture of products which require the use of energy to make them function. The programme of study is designed to enable pupils to acquire knowledge, understanding and skills which can be used in the design and manufacture of products which are meaningful to them.

Technology and Design is essentially a practical subject, therefore, it is anticipated that most learning will take place within the context of practical designing and manufacturing. [...] Throughout this work, pupils should have opportunities to apply skills, concepts and knowledge taught in Science and other relevant subject areas. Technology and Design activities should provide opportunities for pupils to experience the sense of satisfaction and enjoyment that comes from engaging in purposeful activities leading to the production of completed products. Pupils should develop technology and design capability through work associated with four elements: designing; communicating; manufacturing; and using energy and control (DENI, 1992, p3).

For pupils aged 14 to 16 years (known as Key stage 4), the specification in the areas of Designing, Communicating, Manufacturing, and Using Energy and Control is developed. It is the requirement to use energy and control devices that gives the Northern Ireland Technology curriculum much less of a *craft* feel than is prevalent in the rest of the UK. For example:

Using Energy and Control–Knowledge, understanding and skills related to the use of control systems in the realised product. This should include:
- input, process and output;
- system function, performance and feedback;
- one or more of the following control systems
  - computer/microprocessor control,
  - electrical/electronic control,
  - mechanical control,
  - pneumatic control (DENI, 1992, p16).

This curriculum specification in Northern Ireland sets a relatively clear boundary to the subject knowledge (in terms of using materials and processes) a teacher of Technology and Design needs to know. This is very different to the much wider breadth of teacher subject knowledge that is required for Technology, and which is termed Design and Technology (D&T), in England and Wales.

Like Northern Ireland, the National Curriculum is compulsory in all schools in England and Wales. When Technology as a subject was first published for these two countries in 1990, it was configured so that a number of existing school staff to could contribute to it – chiefly teachers of Craft and Design, and Domestic Science (particularly those teachers who taught about Textiles and Food). It was described as a 'subject designed by committee' (McCormick 1994), and as might be imagined, a number of teachers felt that their considerable expertise was being devalued. This was particularly true for the teachers of Domestic Science. In Northern Ireland and Scotland, Domestic Science is a separate subject, but in England and Wales, these material areas were brought into Design and Technology as Food Technology and Textiles Technology, with an entirely different emphasis and purpose to Home Economics, along with the topic of Systems and Control, which is, an area of technology rather similar to that found in Northern Ireland. Clearly D&T has a very wide curriculum base, and in a similar way to Science, cannot be taught to a high level in all its aspects by just one teacher. The latest version of the National Curriculum for D&T in England states:

> Design and technology prepares pupils to participate in tomorrow's rapidly changing technologies. They learn to think and intervene creatively to improve quality of life. The subject calls for pupils to become autonomous and creative problem solvers, as individuals and members of a team. They must look for needs, wants and opportunities and respond to them by developing a range of ideas and making products and systems. They combine practical skills with an understanding of aesthetics, social and environmental issues, function and industrial practices. As they do so, they reflect on and evaluate present and past design and technology, its uses and effects. Through design and technology, all pupils can become discriminating and informed users of products, and become innovators (DfES, 1999 p 1).

D&T is compulsory for all pupils aged from five to 14 years in England and Wales. In England, the subject must be offered on the school timetable until the end of formal schooling at 16 years old; however, not all aspects of D&T are required, and from September 2004 it is no longer compulsory that a pupil takes the subject. This is due to general changes in the 14–19 curricula that have introduced more choice.

In primary schools, teachers required help with the introduction of Design and Technology, especially with the notion of linking designing to making. The Design and Technology Association (DATA), and the Technology teachers' subject association boosted this by producing suggested schemes of work. However, primary teachers identified much more with the purpose of a technological curriculum and its enhancement of manual skills than was the case with the compulsion to teach science.

The Scottish Curriculum for pupils aged five to 14 years attempts to group the learning into broad areas. Unlike the rest of the UK, the school curriculum is not prescribed by statute. Technology sits under the heading of Environmental Studies, along with Society and Science, and has Technological Capability defined as, "understanding appropriate concepts and processes; the ability to apply knowledge and skills by thinking and acting confidently, imaginatively, creatively and with sensitivity; the ability to evaluate technological activities, artefacts and systems critically and constructively" (SCCC,1996, p7). It has as its central purpose:

> The desired outcome of study in technology is technological capability. To help assess pupils' progress and plan suitable experiences, strands and targets have been identified for knowledge and understanding, skills and attitudes. However it is important to recognise that in most technological activities aspects of all of these come into play. Although valuable in its own right, the central purpose of knowledge and understanding in technology is to enable skills to be deployed effectively and attitudes to be well informed. For example, it is necessary to know about materials such as food, textiles and constructional materials and their properties such as softness, durability and strength to be able to decide which to use or to evaluate existing products (Section 5.16 Environmental Studies).

The suggested curriculum is very detailed, as much of it will be taught by general class teachers in primary schools. It states that the three strands of knowledge and understanding required by pupils are about *needs, resources* and *processes*.

> To be good at designing and making, it is necessary to know about all of these. One strand is concerned with understanding how human and other needs can be met by making things or changing the environment in other ways. Another strand is about the resources that can be used to solve problems in technology: tools and other equipment, materials, processes including graphical media and ICT, mechanical and electrical components, time and ideas. A third strand is about the

processes such as drawing, modelling, shaping, constructing, mixing, cutting and controlling, which are used to produce and communicate ideas and solutions to technological problems (Section 5.19 Environmental Studies).

The Craft education roots here are plain. In the upper secondary school, the Technological Activities and Applications mode of the middle school curriculum leads to school leaving examinations and these examinations strongly influence the type of technology education that the pupils receive. Here Craft and Design is most prevalent, and there is no examination in subjects with the word *Technology* in their title. However, courses in Craft and Design (C&D), Graphic Communication (GC) and Technological Studies (TS) are available. Technological Studies is very similar to the curriculum content of Design and Technology: Systems and Control in England and Wales, and Technology and Design in Northern Ireland, and is by far the least popular of the three.

HISTORY

From the early 1970s, Technology as a problem-solving process embracing topics such as electronics, pneumatics and mechanisms, and using construction kits such as Lego and Meccano, was commonly taught in the later/upper years of the larger secondary schools. This rather 'applied science' view of the subject, appealing strongly to males, was promulgated vigorously in the early 1980s with funding from the Department of Employment as part of the Thatcher Government's drive to make the school curriculum more vocationally orientated: the so called Technical and Vocational Education Initiative (TVEI). In-service training was given to teachers of Science and of Craft (usually drawn from the same school), in order to continue the more general development of both school subjects: Craft to embrace new aspects of designing and Science to have a better connection with scientific contexts outside the school laboratory.

In England and Wales a compulsory National Curriculum was introduced in 1990, following a consultation exercise. The final published curriculum document was much more focused than the initial consultative report on Technology as a development of Craft, Design and Technology (CDT) with new areas such as Food Technology and textiles as 'material areas'. More significantly, however, the focus was on Technology as a process. It had four attainment targets:

- Attainment target 1 – Identifying needs and opportunities
- Attainment target 2 – Generating a design
- Attainment target 3 – Planning and making
- Attainment target 4 – Evaluating

This process-based curriculum was difficult to implement as it was also suggested that a wide range of teachers become involved. Implications for initial teacher education were a new need for teachers to offer a breadth of subject knowledge, but also as a specialist area.

After only two years, The Engineering Council produced a damning report by Smithers and Robinson which declared that "Technology in the National Curriculum is a mess" and that the process-based format of the curriculum was not sufficient (Smithers & Robinson, 1992). In 1995 a new version of the curriculum for England and Wales gave a clearer direction to Design and Technology, and the main pedagogical strategies that should be employed:

> Pupils should be given opportunities to develop their Design and Technology capability through:
> Assignments in which they design and make products, focussing on different contexts and materials and making use of:
> - Resistant Materials;
> - Compliant materials and/or food (DMAs- Design and Make Assignments).
>
> Focused practical tasks (FPTs) in which they develop and practise particular skills and knowledge;
>
> Activities in which they investigate, disassemble and evaluate familiar products and applications are also known as IDEAS (National Curriculum DFE/WO, 1995, p6).

Around this time, the Teacher Training Agency (TTA) for England was interested in regulating the standards achieved by teachers entering the profession by specifying for all subjects what competences a teacher should be able to demonstrate. In addition to teaching standards, the TTA also specified the subject knowledge needed to be able to satisfactorily teach the core subjects of English (mother tongue), Mathematics and Science. The subject association for Design and Technology (DATA) thought this level of specificity should apply to Design and Technology too (see figure 1). So from 1994/5, the curriculum content in schools and in teacher training institutions in England was very much clearer.

A consequence of a prescribed curriculum is that there is a danger that ideas and methods stagnate. In 2000, the school Design and Technology curriculum was again revised to give just one attainment target (Design and Technology) and a new and much simplified standards document for initial teacher education courses was introduced in 2002. Consequently, in 2003 new subject knowledge expectations for teachers entering the profession were suggested by DATA. The revised school curriculum strongly promoted new technologies – in particular the use of CAD/CAM software packages.

In Northern Ireland, the more restricted subject knowledge encompassed by Technology and Design has meant that the subject has been more stable. Technology and Design is intended:

"to enable all pupils to become confident and responsible in solving real life problems, striving for creative solutions, independent learning, product excellence and social consciousness" [Technology and Design Ministerial Report 1991, p15].

And in the current syllabus for the leaving examination in Northern Ireland, it is suggested that all candidates should have a working knowledge and understanding of:

- Designing;
- Communicating;
- Manufacturing;
- Using Energy and Control *(CCEA, 2000)*.

Scotland saw a similar interest in the development of Technology in terms of structures, mechanisms, electronics and so forth in the rise of the subject Technological Studies in 1988. As in England, this was promoted by funding available under the Technical and Vocational Education Initiative (TVEI). However, in Scotland the craft tradition continued on a parallel track as there was no unification of contributory subject areas in quite the same way as happened under the English National Curriculum. According to the entries for the school leaving examination, Technological

Studies numbers fell from 6,076 in 1994 to 3,649 in 1999, yet in the same period Craft and Design rose from 11,649 to 13,783 and Graphical Communication rose from 5,778 to 7,319 (Dakers and Doherty, 2002). Here clearly, pupils were voting with their feet and moving away from the perceived difficult topics in favour of the more traditional craft areas.

OVERVIEW OF TECHNOLOGY TEACHER EDUCATION

The breadth of school technology is such that many technology teachers who were traditionally teachers of Craft, Domestic Science or Science do not have the subject knowledge background, even in their specialist area, to teach the entire current school Technology curriculum with confidence. In 1994, DATA surveyed the training needs of D&T teachers in England and Wales (DATA, 1994). That survey assumed that the most important issue to address was the particular subject elements in the national curriculum for D&T in which teachers were either untrained, or considered themselves to be inadequately trained to teach. A given assumption was that only the 'rectification' of the teachers' subject knowledge 'deficit' was required. New pedagogical strategies still have a relatively low priority for D&T teachers. This is very different from teachers of History or Science where the common expectation is that in-service courses will lead to novel pedagogy such as techniques for promoting discussion or group solutions to problems.

This emphasis on skill or subject knowledge rectification has had a profound impact on pre-service technology teacher education. Many teacher preparation courses, even those of just one-year duration which follow a degree in a technological subject, might spend half or more of the time introducing the student teachers to the subject content of school technology. This 'school knowledge' would include for example expertise in general instruction in the particular CAD/CAM packages to be found in schools as well as more focused subject development to 'plug the gaps' identified on a personal subject audit.

In 1995, the subject association for Design and Technology in England and Wales set out for England some subject knowledge guidelines for common areas of Designing and Graphical Communication, and for specialist areas of Resistant Materials, Systems & Control, Food Technology, and Textiles Technology. This was revised in 2003. A secondary student teacher is expected to teach two of these material areas adequately to pupils aged 14 years (key Stage 3) and one specialist area to students of 18years (Key stage 4 and beyond). The diagram in Figure 1 sets out the current subject knowledge expectations.

Figure 1. DATA Minimum subject competence areas

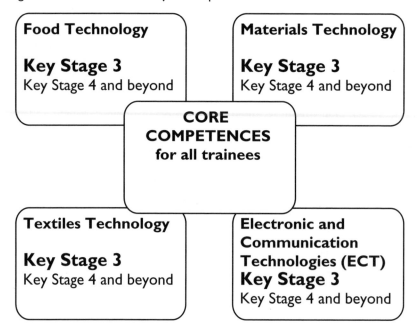

All newly qualified teachers should be able to exhibit the Core Competences such as:

C.1   understand and use a range of strategies and approaches to identify and clarify design problems (for example, analyse an existing product with the view of improving it; observe a product being used with the view of improving its performance for the user; use a modern or smart material as a starting point for a new product – such as an EL panel);

C.3   research a wide range of information sources appropriate to the problem. Analyse and select this information to inform their designing (for example, research an art and design movement, conduct taste testing, seek the views of an end user/expert);

C.7   use techniques, processes and procedures appropriate for each of the specialist fields to manufacture products and systems;

C.8   when planning and conducting design and technological activities give due regard to Health and Safety of their pupils, themselves and other adults and be aware of current, relevant Health and Safety responsibilities, legislation and liability.

C.10   an awareness of industrial methods and approaches to design, manufacture and quality control on production (for example, batch, mass and continuous flow production);

C.11   have an understanding of citizenship and values and how these relate to Design and Technology (for example, sustainable design such as recycling materials, designing products with dual functions, GM foods, alternative/renewable energy sources, fair trading).

C.12   critically use ICT to enhance teaching and learning in Design and Technology using modern technologies (for example, Interactive whiteboards, scanners, digital cameras, and presentation applications).

C.15   nurture a creative teaching and learning environment where pupils feel confident and safe to experiment, explore and take risks (DATA, 2003).

Although following these DATA guidelines is not a formal requirement, virtually all of the teacher education institutions have endorsed this subject knowledge framework from the subject association, although they rarely take on student teachers in all four material areas.

Two words in Figure 1, *trainee and competences*, highlight some other changes that have impacted on technology teacher education in recent years in all nations except Scotland. First there was a move to reduce the time spent in Higher Education and increase the time spent in school. The use of the word 'trainee' rather than 'student' emphasizes this work-based training view of teacher preparation, and that has increased in its stridency in recent years. Together with the increase in school-based work was an assessment of teacher quality to enter the profession, based on a series of competence statements. A teacher who can exhibit all the competencies (and early versions of the list of competences were very detailed indeed) can pass and receive Qualified Teacher Status; those who do not, fail. This, in itself, is not necessarily a retrograde step. On the contrary, a clear and common understanding of the facets of good teaching that need to be considered when assessing a novice teacher, made explicit and shared between student (trainee), school mentor and the University tutor is to be welcomed. But the move was strongly criticized for introducing a 'technicist' view of the role of a teacher, and is not in line with the more comfortable notion of a teacher as a reflective practitioner.

In Scotland, Higher Education resisted this move to school based training, although a set of agreed competency areas was introduced. A major stumbling block to whole scale policy reform resulted from the reluctance of school teachers to become mentors with the extra work load required. In Northern Ireland, the competency model was introduced with refinements. There, instead of a list of aspects of teaching all of which needs to be demonstrated *before* entry to the profession, the competencies are also used during the initial induction years; some competencies are stressed in pre-service, others during the first year of employment.

STRUCTURE OF TECHNOLOGY TEACHER EDUCATION

There is a great diversity of routes into teaching in England and Wales. The rationale is: a common set of detailed standards for acceptable performance as a teacher on entry to the profession, and a similar commonly agreed catalogue of appropriate subject knowledge, after which the way those standards are achieved is irrelevant. Indeed, if a person can already meet the requirements and has the teaching subject knowledge, they may not need any pre-service education, but merely an assessment to demonstrate proficiency. However, there are two main models of teacher training in England, Wales and Northern Ireland:

1. The *concurrent models* of Bachelor of Arts in Education, the BA(Ed), or Bachelor of Education (B.Ed) are most common for primary teachers where subject study and pedagogy are studied side by side. These are either three-year or four-year courses -or exceptionally, two year courses for students with some Technology background.

2. The alternative *consecutive model* leads to the Postgraduate Certificate in Education (PGCE), traditionally one year and virtually just pedagogy with relatively little subject knowledge included, following a subject specific degree. Most secondary teachers are prepared for teaching via the PGCE which is commonly full-time and 36 weeks in duration, 24 weeks of which are spent on school experience. However, due to the wide range of 'school knowledge' needed to teach such an eclectic subject as D&T, even the one year course often includes

modules for subject enhancement. This is less prevalent in Scotland and Northern Ireland where the school technology curriculum is more focused, but here too the specifics of equipment such as pneumatics and electronics kits and common software for control and pupil-friendly CAD/CAM, occupy a major part of the course. Also PGCE students in Scotland and Northern Ireland are required to follow an additional practical workshop course in Health and Safety matters, lasting about six weeks.

New models of teacher training involving the modularisation of programmes were introduced in 2000, and, in England and Wales, there are new employment-based routes where a graduate is employed and receives on-the-job training. In December 1999, the Minister for Schools announced:

"Part of that process is to introduce new, flexible routes into the profession. I am today announcing new teacher training targets that have enabled the Teacher Training Agency to accept bids to bring on stream 660 new modular places from September 2000. [...] As well as modular provision, we have introduced the Graduate and Registered Teacher Programmes, which allow people to earn a salary while they train. Some 850 people have already achieved or are working towards qualified teacher status in this way. Many would not have found it easy to follow traditional courses. In the new year I expect to announce proposals to provide for at least a doubling of the numbers going through these employment-based routes" (Morris, 1999).

Variations on the PGCE model include two-year full time courses for students whose degree profile is not quite so well matched to the school subject and, following the Morris announcement, 'flexible' versions were introduced, which can be studied part-time and tailored to the students' personal needs. For example, a mature student might want a long part-time course where study and school experience can fit around home and employment commitments. Conversely a student who is a school technician may already be very familiar with school technology. Such a student would not need to study a complete course.

In England and Wales, Universities and Colleges of Education are not the only sites for teacher education. In addition to the employment routes, there is also School-Centred Initial Teacher Training (SCITT) which might offer a PGCE validated by a University.

In Scotland, there are three routes to enter the teaching profession:
- a four-year BEd for primary teaching and for some specialist secondary teaching such as technological education (and also music and physical education). For example the University of Glasgow offers a Bachelor of Technological Education (BTechEd);
- a combined degree including school experience, study of technology and study of education;
- a postgraduate certificate in primary or secondary education (PGCE) following a degree. By 2006 the PGCE in Scotland will be changed to the PGDE (Professional Graduate Diploma in Education) to ensure consistency within the Scottish Credit and Qualifications Framework. However, there is a general desire to move away from a PGCE as a route into teaching for Technology teachers due to the breadth of subject knowledge required. Tables 1 and 2 provide data on the number of students (trainees) who begin different routes into technology teaching for each nation state. Tables 3 and 4 indicate the duration of each route. Note the variation in the length of courses available even when they lead to the same qualification.

Table 1.
Numbers of Secondary Student Teachers on Postgraduate Technological/Technical Education Courses (PGCE)

Year	2001	2002	2003	@ June 2004
England	737	785	908	668
NI	11	11	11	11
Scotland	n/a	50	44	66
Wales	71	63	80	69

Source: GTTR & TTA

Table 2.
Numbers of Student Secondary Teachers
starting different types of Courses

	2001	2002	2003
Undergraduate England	232	170	170
Undergraduate Wales 5	9	67	45
Undergraduate Scotland	36	36	48
Undergraduate NI	26	26	26
GTP (Employment Route-England)	143	193	230
RTP (Employment Route for non-Graduates–England)	11	14	7
Overseas Trained Teacher Route (England)	34	35	46

Source: TTA/HEFCW/SEED

Table 3.
Number of Undergraduate Courses in Teaching Technology

	England			NI			Scotland			Wales		
Duration in Years	2	3	4	2	3	4	2	3	4	2	3	4
Primary										1		
Primary/Secondary (KS2/3)		2										
Secondary	7	5	2			2	1(ord)		2	2	2	

Source: UCAS

Table 4.
Number of Postgraduate Courses in Teaching Technology

	England		NI	Scotland	Wales
Duration in Years	1	2	1	1	1
Primary	2				
Primary/Secondary (KS2/3)	2	1			
Secondary	49	7	1	4	2

Source: DENI/GTTR/SEED

Key: DENI – Department of Education Northern Ireland
GTTR –Graduate Teacher Training Registry (England & Wales)
HEFCW – Higher Education Funding Council Wales
UCAS - Universities and Colleges Admissions Service (UK)
SEED – Scottish Executive Education Department
TTA – Teacher training Agency (England)

A shortage of Technology teachers exists across almost all of the United Kingdom. In England, in 2000 for example, Technology under-recruited for initial teacher preparation courses by 41%, the largest deficit of any subject, but by 2003 the situation was improving, with a 26% increase over the 2002 level. A number of financial inducements have helped; across the UK all postgraduate students receive free tuition on teacher preparation courses plus a £6,000 bursary while training in England and Wales, and for those beginning their teaching in England a 'Golden Hello' of £4,000.

Nature of Study

Since the mid-1980s there has been a desire on the part of government to improve the quality of initial teacher preparation courses. This is true of all UK countries but it took on different forms; being most regulated in England and least in Scotland. In England, Wales and Northern Ireland, however there was general acceptance that one way to gain uniformity across the diverse teacher education sector was to move to a Competence Model of assessment. All teachers, regardless of their teacher education route into the profession, must satisfy the same standards by exhibiting the same competencies. This became a formal requirement in the 1990s and, in many ways, provided a framework which has led to considerable uniformity in the nature of study. For example, in England, all student teachers undertake a practicum in at least two schools under a partnership arrangement between higher education and their placement schools. The Universities provide a tutor to teach aspects of pedagogy while the students are in college and the schools provide a mentor to help support, coach and assess the student teacher in the classroom. The enhanced status of practical teaching, together with a government-required standards framework has moved initial teacher education towards a rather functional view of the teacher and away from many of the traditional disciplines of education courses such as philosophy, sociology and psychology. In its place are lectures, seminars and workshops designed to ensure all student teachers (or 'trainees' as they are referred to by English government documentation) can show evidence of appropriate standards in the following areas:

1. Professional Values and Practice
2. Knowledge and Understanding
3. Teaching
   3.1 Planning, expectations and targets
   3.2 Monitoring and Assessment
   3.3 Teaching and class management (TTA 2002)

This order does not suggest a hierarchy of importance. All standards or competencies in all areas must be achieved. Almost all secondary Technology courses in England and Wales have agreed to implement the requited Government standards for teachers' Knowledge and Understanding in the same way as suggested by DATA and shown in Figure 1.

Wales, Northern Ireland and Scotland take a similar competences-based approach. For example, in Scotland the Scottish Office Education and Industry Department (SOEID) suggested:

1. Competences relating to Subject and Content of Teaching
2. Competences relation to the classroom
   2.1 Communication and approaches to teaching and learning
   2.2 Class organisation and management
   2.3 Assessment
3. Competences relating to the school and the education system
4. The values, attributes and abilities integral to professionalism.

AN EXAMPLE: SHEFFIELD HALLAM UNIVERSITY

A suite of courses which illustrate very clearly the consequences of a competency-based, standards driven approach to the education and training of Technology teachers is that offered by Sheffield Hallam University in England (see Table 5). This University has taken a radical and innovative approach to the question of the shortage of teacher supply in the UK by taking account, in a flexible and pragmatic manner, of firstly, what people bring to a course and consequently, what they need to study in order to achieve the required final standards for Qualified Teacher Status (QTS). In many ways, it is irrelevant whether a course is labelled undergraduate or post-graduate. What is more important is what entrants know, understand and are able to do at the start of a programme and the time they therefore need to achieve QTS. The Sheffield Hallam University Centre for Design and Technology Education offers the following courses:

- 3 year BSc (Hons) Design and Technology with Education and Qualified Teacher Status (QTS)
- 2 year BSc (Hons) Design and Technology with Education and Qualified Teacher Status (QTS) and 2 year Design and Technology PGCE with Qualified Teacher Status (QTS) – these two awards follow the same contributing courses
- 1 year PGCE – This one-year course is also the final year of all the programmes and is also offered in a part-time flexible version

It may seem unusual that a post-graduate award (PGCE) is linked so closely to undergraduate study. This is very common indeed in the UK (see Tables 1 to 4) where a Postgraduate Certificate in Education is considered to be postgraduate in time, but not level. The overlapping nature of these courses is shown by the following list of modules at Sheffield Hallam University:

Table 5.
D&T Initial Teacher Education Programme – Sheffield Hallam University

Credit Points	3-year BSc	2 Year BSc and 2 year PGCE	1 Year PGCE and 'flexible' PGCE
	Year 1		
30	Foundation design Studies		
40	Foundation manufacturing Studies		
30	Foundation technology studies		
20	Introduction to D&T in secondary schools		
20	Foundation design and manufacture project		
20	Professional Standards A		
	Year 2	Year 1	
20	Design Studies	Design Studies	
20	Manufacturing Studies	Manufacturing Studies	
20	Technology Studies	Technology Studies	
40	Design and Manufacture Project	Design and Manufacture Project	
20	Understanding D&T in Context	Understanding D&T in Context	
20		D&T in Secondary Schools	
20	*Developing as a teacher*	Resources for teaching and learning in D&T	
20	Professional Standards B	Professional Standards B	
Professional Year	Year 3	Year 2	Year 1
30	Teaching and Learning in D&T	Teaching and Learning in D&T	Teaching and Learning in D&T
10	Developing professional skills and knowledge in D&T	Developing professional skills and knowledge in D&T	Developing professional skills and knowledge in D&T
20	Current issues in D&T education	Current issues in D&T education	Current issues in D&T education
20	Developing D&T practice	Developing D&T practice	Developing D&T practice
20	Needs analysis/ teaching placement	Needs analysis/ teaching placement	Needs analysis/ teaching placement
20	General professional studies	General professional studies	General professional studies

Note the few courses (shown in italics) that are unique to a particular year-group.

Without going into too much detail, it is possible to see progression across the years (naturally Year 2 courses are generally more demanding than Year 1) and to estimate relative weight of study, as 20 points is twice the work of 10 points and so forth.

As Table 5 clearly illustrates, there is a common final professional year and a large degree of commonality between the second year of the three-year programme and the first year of the two-year programme. The courses are designed so that students are given sufficient teaching and time to study as a common starting point for their professional year – which is the same as the one-year PGCE. Given the relatively large number of students enrolling for D&T courses – and Sheffield Hallam University is one of the biggest providers – this 'funnelling' of students from year 1 to a bigger year 2 and then a common final professional year is extremely efficient in terms of staffing, equipment and accommodation.

A common thread running through all courses at Sheffield, is the strong commitment to developing a student teacher's subject knowledge and skills. Unusually, D&T Teacher Education lecturers at this University are not located in a Faculty of Education, but in the Faculty of Arts, Computing, Engineering and Sciences. Staff are certain that this location has had a profound influence on the nature of their course, not least of which is the ease with which students are able to develop subject expertise through access to high quality workshops and equipment. The main subject knowledge required is in Resistant Materials, and Electronics and Communications Technology (often called Systems and Control) although a Food Technology route is available as a one year PGCE. For example, even in the professional year with its limited 12 weeks in university, the 10 point module Developing Professional Skills and Knowledge in Design and Technology aims to enable the auditing of subject knowledge and to facilitate subsequent reflection. As stated in its learning objectives, students will be able to:

- demonstrate sound knowledge and understanding of D&T and its creative application in the contemporary educational context,
- demonstrate specialist skills, knowledge and understanding in the field of design and technology,
- produce an e-portfolio using appropriate software and peripherals,

- communicate information and ideas clearly in two and three dimensional visual formats appropriate for different audiences.

This module also enables students to develop skills in communication and ICT.

Another module, Developing Design and Technology Practice, emphasises the links between subject content and pedagogy, and here the intended learning outcomes suggest a student will be able to:

- demonstrate a working knowledge of materials, resources and processes appropriate for teaching design and technology in a second, or additional field, to the required level within a field of D&T;
- have developed specific teaching and learning strategies appropriate for teaching design and technology in a second or additional field to the required level;
- be aware of the need for and the use of purpose-made educational aids and technologies as appropriate within a given field of D&T.

The links to the common DATA subject knowledge framework in Figure 1 are plain.

As a final example, the module Current issues in D&T Education is noteworthy. It provides an opportunity for the students to use their particular expertise to become agents for change in the profession. This module helps the students look forward to their first teaching post. It also aims to enable them to utilise current issues for their future practice.

In common with many teacher education courses in secondary Technology, virtually all aspects of teacher preparation are highly subject specific. However, there is one course of lectures and seminars addressing General Professional Studies (taught in the School of Education) which looks at broader professional practice and links this to school-based experience. Here the values required by teachers and the legal requirements, for example, would be explored.

TEACHER CERTIFICATION

In recent years, General Teaching Councils (GTCs) have been established in England, Wales and Northern Ireland. Scotland has had such a Council for much longer and it has a considerable voice in the development of the profession there. The General Teaching Councils have two main functions: to advise the respective policy makers in the different regional governments on education policy, reflecting the views of teachers, and to regulate the profession to guarantee and maintain professional standards. These bodies, therefore, now admit teachers to the profession and formally confirm the Qualified Teacher Status.

Universities and other Designated Recommended Bodies which have been approved by the Teacher Training Agency in England, the GTC in Scotland and the regional governments in Wales and Northern Ireland make recommendations to respective GTCs that graduates of their programmes should become teachers. They are therefore able to teach on the qualified teacher scale. People who have QTS in England and Wales can teach in any sector – primary or secondary – irrespective of the course that was followed. It is possible to teach in most areas of the UK without a teaching qualification provided criminal record checks have been passed. Indeed, in some urban areas, technology teachers are in such short supply that unqualified teachers are paid at the qualified rate.

In Scotland, every teacher graduating from an initial teacher preparation course is guaranteed a post for their first year. Here, in common with other areas of the UK, they will undergo a period of induction into the profession and probation to confirm their qualified status in the profession. In England, for example, newly qualified teachers complete an induction period of three terms or equivalent, beginning when they first take up a post that lasts for a term or more. They are entitled, by law, to an Induction Support Programme that combines two interrelated and equally important aspects:

- an individualised programme of monitoring and support; and
- an assessment of the Newly Qualified Teacher's (NQT's) performance.

Near the end of their induction period, new entrants to the profession are assessed against a set of Induction Standards. These include a requirement to continue meeting the Standards for the award of QTS consistently and with increasing professional competence, and to progress further in specific areas. If they fail to reach such standards they will not remain registered with the General Teaching Council for England, and cease to be allowed to continue teaching in a state-maintained school or non-maintained special school. The private sector remains outside these regulations, including the requirement to have QTS, although most of the more prestigious private schools recognise the quality implied by such a qualification.

CONCLUSION

In 1990, England and Wales took the lead to make Technology a compulsory subject, part of all pupils' general education between the ages of five and 16 years. Today, some of that compulsion has weakened, but it is still there for all pupils – in all areas of the UK – from the ages of five to 14 years. It is also a key area in the evolving curriculum for the 14 to 19 age range as it has a clear link to vocational courses. Although the subject domain is very wide in England and Wales, early feuding between the different contributory subjects has largely quietened down and common links in terms of design methodology have been strengthened.

It has been claimed that once someone comes up with a way of preparing teachers, in schools or Higher Education, on the job, or on Undergraduate or Postgraduate courses of various lengths, full-time, part-time or by open and distance learning – someone else, somewhere in the UK, has already done it! This pragmatism is based on a utilitarian, craft view of teachers and teaching which goes back to Victorian times, and which is strengthened by competence-based assessment arrangements and tight government regulation; whatever the route into the profession, the entry standards must be the same.

The consequence of these different training and education routes and common approaches, combined with financial incentives for training and staying in the profession, have had a beneficial impact on the number and quality of people now wishing to become teachers of Technology in the United Kingdom. Technology teachers are still in short supply, but the worse fears have been allayed.

REFERENCES

Common Curriculum Examination Authority (CCEA) (2000) *Specification for GCSE in Technology and Design – From September 2001*, Belfast.

Dakers, J. R. and Doherty, R. (2003) 'Technology Education' in T.G.K. Bryce and W.M.Humes (Eds) *Scottish Education, Second Edition Post-Devolution*. Edinburgh: Edinburgh University Press.

Design and Technology Association (DATA) (1994) *A survey of the Qualifications and Training Needs of D&T Teachers in Secondary Schools*. Research Paper No. 2. Wellesbourne, DATA.

Design and Technology Association (DATA) (2003) DATA Research Paper 4 – 'Minimum Competences for Trainees to Teach Design and Technology in Secondary Schools'. Wellesbourne: Design and Technology Association. (See http://web.data.org.uk/data/index.php)

Department for Education and Skills(DfES) (1999) *National Curriculum: Design and Technology*. London: http://www.nc.uk.net/index.html (Accessed October 2004)

Department of Education Northern Ireland (DENI) (1991) *Proposals for Technology and Design in the Northern Ireland Curriculum*. Report of the Ministerial Technology and Design Working Group, July 1991

Department of Education Northern Ireland (DENI) (1992) *Northern Ireland Curriculum : Technology & Design Key Stage 3* http://www.deni.gov.uk/parents/key_stages/d_key_stages.htm (Accessed October 2004)

Department for Education/ Welsh Office (1995) *Technology in the National Curriculum*. London: HMSO.

McCormick, R. (1994) 'The coming of technology education in England and Wales' in F.Banks (ed) *Teaching Technology*. London: Routledge.

Morris, E. (1999) *Ministerial Press Release*. Department for Education, December, 1999.

NCC (1990) Non-Statutory Guidance Technology. York: HMSO.

SCCC (1996) *Technology Education in Scottish Schools: A Statement of Position from Scottish CCC*. Dundee: SCCC.

Scottish Office Education and Industry Department (SOEID) (1999) *Competences in School Experience*. Edinburgh.

Smithers, A. & Robinson, P. (1992) *Technology in the National Curriculum: Getting it Right*. London: The Engineering Council

TTA (2002) *Qualifying to Teach: Professional Standards for Qualified Teacher Status and Requirements for Initial Teacher Training*. London: Teacher Training Agency.

Technology Teacher Education in the United States

Chapter
12

Mark Sanders
Virginia Polytechnic Institute & State University, USA

INTRODUCTION

In contrast to education in many countries, public schooling in the United States (US) is a matter left largely to the states and local school divisions. The funding pattern for public education underscores this fact. Only about 7% of school funding derives from the federal government, while the remaining 93% is split somewhat evenly between state and local governments. Thus, most decisions about what is taught and how it is taught are made by state and local decision-makers. There is no such thing as a 'national curriculum' in the US, and teacher licensure regulations are established independently – and therefore somewhat differently – in each of the 50 states.

Over the past two decades, educational reform efforts have resulted in both national and state standards in all of the 'academic' subject areas, including mathematics, science, social studies, and language arts. Nationally developed standards, including the *Standards for Technological Literacy* (STL, ITEA, 2000) are essentially a set of recommendations developed and championed by professionals within school subject disciplines, working in cooperation with their professional associations. Individual states and localities are at liberty to determine the extent to which they incorporate any or all of the ideas embedded within those nationally developed school subject standards. So, while the STL have been well-received by the profession and will influence the field in many ways over the coming decades, they remain a set of recommendations rather than a set of required content standards.

In contrast to the recommended nationally developed standards, nearly all states have mandated standards in the academic subject areas, including English, mathematics, science, and social studies. Spurred by a nationwide 'accountability' movement, states are requiring all students to take statewide assessments in these academic subject areas. This is causing local school divisions to focus resources on the academic subjects, potentially to the detriment of elective subjects such as Technology

Education (TE). Technology educators in several states have been successful in getting language which addresses the study of technology incorporated into their state standards. But even in these states, it has not yet resulted in statewide mandatory enrolment in Technology Education courses beyond a relatively brief middle school experience. Some local school divisions have countered the state standards movement with publications that claim specific Technology Education learning experiences help students achieve specific state standards–such as those in mathematics and science. But here again, that strategy has not resulted in compulsory Technology Education courses.

Many state departments of education and some local school divisions employ Technology Education teachers, teacher educators, and/or curriculum specialists to develop curriculum guides. These developers have the option of using the STL to guide them in this work, but once again, this is voluntary. In 1990, the International Technology Education Association established a Center to Advance the Teaching of Technology and Science (CATTS) to develop curriculum materials based upon the STL. Currently, just 12 of the 50 state departments of education hold an annual subscription to CATTS, which allows these states to distribute CATTS publications developed during their subscription year to Technology Education teachers throughout their states–this is also very different from a 'national curriculum.' In many states, Technology Education is administered under the umbrella of vocational education (Career and Technical Education, CTE), for historical reasons explained later in this paper. This administrative practice has often influenced decisions with respect to Technology Education curriculum and teacher licensure.

For all of the aforementioned reasons, Technology Education teacher licensure, curriculum, and practice in the US vary significantly from state to state and from one local school division to another. Despite these differences and the various efforts to infuse Technology Education into the school curriculum, the average student in the US currently gets only a brief exposure, if any, to Technology Education throughout 12 years of compulsory education.

The Structure of Technology Education in the US

Technology Education as an Elective Subject in Grades 6–12

Technology Education is, for the most part, an elective (optional) subject in grades 6–12. The primary exception to the elective nature of Technology Education courses occurs at the middle school level (grades 6–8), where in many localities, all students are required to enrol in a Technology Education course–though typically only for 6–18 weeks in duration. These courses generally introduce students to a wide range to technologies, with course titles such as Introduction to Technology, Inventions and Innovations, or Technological Systems. Due largely to the impact of digital technologies and entrepreneurship, many of the 'general laboratories' of the 1970s have been replaced by 'modular laboratories.' These typically consist of 6–15 modules, each of which provides students working in pairs with an activity representing the different technological systems (for example, information and communication, transportation, power and energy, manufacturing, construction, medical, and agriculture and bio-related technologies).

At the high school level, Technology Education is an elective subject. Although a few states, such as Maryland, lay claim to a state-legislated Technology Education requirement, there are far too few Technology Education teachers to deliver on this mandate. In the few states in which this state-legislation has occurred, courses other than Technology Education are routinely substituted for Technology Education, to comply with the requirement. Several states, such as New York and Massachusetts, are attempting to integrate technology standards with science and mathematics standards. In practice, however, the very limited number of Technology Education teachers across the K–12 continuum (including virtually none at the elementary grade levels) typically results in other teachers, such as elementary or science, addressing the technological component in very limited ways.

Technology Education in Grades K–5

Where implemented, elementary school Technology Education is generally highly regarded by those closest to the action: teachers, students, parents, and school administrators. But without a public mandate, these successes have been difficult to sustain over time, and no states address Technology Education as a stand-alone subject in grades K–5. There have been efforts to incorporate the study of technology into elementary grades since the early 19th century (see, for example, Battle, 1899; Bonser and Mossman, 1923; Gerbracht & Babcock, 1969; Miller and Boyd, 1970; Scobey, 1968; Winslow, 1922). A relatively small number of individuals–teacher educators, state supervisors, and elementary teachers have kept elementary school Technology Education alive through pre-service teacher education courses, in-service workshops with elementary teachers, and funded curriculum projects. The Technology Education Council for Children, a division of the International Technology Education Association (ITEA) has provided leadership for elementary school Technology Education in the US and is the primary force behind Technology and Children, an ITEA serial publication that focuses solely on elementary school Technology Education. Despite these efforts, elementary school Technology Education remains very sparse in the US.

HISTORY

In the last quarter of the 20th century, Technology Education emerged from Industrial Arts (IA) education, emphasizing different purposes than those championed in the IA era (Sanders, 2001). In the early years of the 20th century, encouraged by the work of John Dewey and the progressive education movement, a growing number of educators believed a general course of study that addressed industry and related social issues would benefit all students as citizens of our democratic society in the industrial age. This perspective was exemplified in Bonser and Mossman's (1923) definition of IA, as "... *a study of the changes made by man in the forms of materials to increase their values, and of the problems of life related to these changes.*" Those who aligned philosophically with Bonser & Mossman's interpretation worked toward different general education goals than those in manual training who espoused a vocational approach to the curriculum. The Smith-Hughes Vocational Education Act of 1917, the first federal funding for any component of public education in the US, supported vocational education as a means of providing a new source of industrial workers in America. This legislation further encouraged the two factions to split along philosophical lines into what became vocational education and IA education. Vocational educators used the federal funding and Smith-Hughes' philosophy to develop trade and industry education and other vocational subject areas for some students, while IA educators continued to promote general education ideals and curriculum for all students.

Following World War II, leaders in the field began to study the idea of a curriculum grounded in the concepts of 'technology' rather than 'industry,' a trend initiated with William E. Warner's 1947 presentation titled *A Curriculum to Reflect Technology* (Warner, Gary, Gerbracht, Gilbert, Lisack, Kleintjes, and Phillips, 1947). In 1985, following four decades of professional dialogue, the American Industrial Arts Association changed its name to the ITEA. Since then, most programs in schools have followed suit (Sanders, 2001), though in practice, there remain widely varying approaches to Technology Education curriculum, content, and method.

While leadership in the profession espoused a general education philosophy, practice continued to focus largely on tool skills into the 1980s (Dugger, Miller, Bame, Pinder, Giles, Young, & Dixon, 1980; Schmitt & Pelley, 1966). This emphasis on tool skills and prevocational goals has always been a source of ambiguity in the profession (Lewis, 1996; Sanders, 2003). Leaders in the late 1960s successfully lobbied for the inclusion of IA in the 1972 Vocational Education Act, and within a few years, fully three quarters of the states were using federal vocational monies to fund some aspects of IA (Steeb, 1976). This trend and the accompanying philosophical ambiguity continue today, with 83% of the responding states reporting the use of monies provided by the federal vocational legislation for Technology Education (Sanders, 2003).

Over the past two decades, the ITEA has steadfastly promoted the new Technology Education agenda. Three ITEA publications have been instrumental in the transition from Industrial Arts to Technology Education: *Conceptual Framework for Technology Education* (Savage and Sterry, 1990); *Rationale and Structure for the Study of Technology;* (ITEA, 1996); and *STL: Content for the Study of Technology* (ITEA, 2000). State departments of education, technology teacher education programs, and local school divisions have begun to use these as they upgrade their curricula.

OVERVIEW OF TECHNOLOGY TEACHER EDUCATION

Technology teacher education (TTE) began to emerge in the late 19th century. In 1886–87, a manual training laboratory/course was established at the State Normal School[1] at Oswego, New York (NY). A year later, NY state legislation established similar manual training teacher education programs in normal schools throughout the state *(Industrial Arts Teacher Education at Oswego to 1941)*. Manual Training/IA teacher education programs subsequently developed at normal schools throughout the US.

Estimates of the number of technology teacher education programs and graduates have historically been made from data culled from the *Industrial Teacher Education (ITE) Directory*. Because the *Directory* now includes many programs that do not prepare teachers, estimating the numbers of technology teacher education programs and graduate requirements is somewhat subjective. To get a reasonable estimate of the current numbers, the data in the 2004–05 *ITE Directory* (Schmidt and Custer, 2004) was reviewed. Institutions that listed Technology Education graduates, Technology Education licensure candidates, faculties identified in the *Directory* as having technology teacher education responsibilities, and/or programs known to be a recent source of Technology Education graduates, even if none of the aforementioned criteria were met, were considered 'active' technology teacher education Programs. Using those criteria, the author identified 70 technology teacher education programs as currently active. This group included at least seven programs the author deemed 'questionably active.'

Historically, there have been four large technology teacher education programs in the US: State University of New York at Oswego, the University of Wisconsin–Stout, Millersville University of Pennsylvania, and California University of Pennsylvania. Anecdotally (and historically), these four institutions have been said to produce about one fourth of the technology teacher education graduates each year. For the 2003–2004 year, these four institutions reported a combined 209 of the total of 550 Technology Education baccalaureate degrees reported (by the 42 institutions who reported Technology Education baccalaureate degrees in the

---

[1]Normal schools were state-funded schools specifically established for public teacher education. The first Normal School was established in Massachusetts in 1839.

2004–2005 *ITE Directory*). That works out to an average of 13.1 Technology Education graduates per institution. The other 38 institutions reporting Technology Education baccalaureate data in the 2004–05 *ITE Directory* (for the 2003–2004 year) accounted for a combined total of 341 graduates (an average of 9.0 Technology Education graduates per institution). Assuming the 28 non-reporting institutions averaged 9.0 graduates per institution (likely an overestimate for those non-reporting institutions), the 70 technology teacher education institutions in the US would have produced an estimated total of 802 Technology Education baccalaureate degrees in 2004. Finally, it is worth noting that there are some states that have no technology teacher education programs.

STRUCTURE OF TECHNOLOGY TEACHER EDUCATION

Most of the estimated 70 technology teacher education programs in the US operate four-year undergraduate baccalaureate degree programs leading to Technology Education licensure. A relatively small, but increasing percentage of Technology Education teachers are prepared through fifth year, masters/licensure, and alternative licensure models described below.[2] Technology teacher education programs are found in all types of four-year post secondary institutions and are housed in a wide range of administrative units, including colleges, schools, or departments of education, arts and sciences, applied science and technology, technological studies, engineering, and human resources.

Technology teacher education in the US consists of three components: general education, pedagogy/professional education, and technical coursework. The general education component is a core of courses that most colleges/universities require of all students, regardless of the field they choose to pursue. These courses are typically arts and science courses in English, mathematics, social science, natural sciences, and the humanities. These general education courses are decided upon by university communities and are taught, for the most part, by faculty in the arts, sciences, and humanities.

---

[2]A significant, yet unknown, percentage of current Technology Education teachers do not hold a valid Technology Education teaching license. They are employed on a "provisional" basis and in theory, at least, must be terminated after 3 years in this status if they do not earn licensure.

The pedagogy/professional education component of the technology teacher education curriculum generally includes courses in educational foundations (for example, the historical, philosophical, and social foundations of education), educational psychology, curriculum development, and instructional methods. Education majors representing all of the subject areas typically enroled in the educational foundations and educational psychology courses concurrently, while the Technology Education faculty commonly teach curriculum and methods courses to Technology Education majors. In addition, the pedagogy/professional education component generally includes early clinical education experiences prior to student teaching, which typically occurs during the final year of the program.

The third component of technology teacher education comprises a wide range of technical courses that provide the technical knowledge and skills needed to be an effective Technology Education teacher. Historically, these courses were taught by IA teacher educators and included technical content typically taught to students in grades 6–12, along with more advanced technical content. Currently, it is more common for these courses to be taught by highly technical faculty working in a non-teaching degree program, such as Industrial Technology, with content that is likely to be more technical than that taught at the secondary school level.

In practice, these three components of teacher education have been delivered in different configurations (Custer and Wright, 2002; Householder, 1993; Zuga, 1997). Following are brief descriptions of the most common technology teacher education models currently implemented in the US.

Technology Teacher Education Models

Traditional Four-Year Technology Teacher Education Model

Until the 1970s, the IA education faculty generally had responsibility for teaching both the in-major pedagogy/professional courses (curriculum, method, and clinical experiences) as well as all of the technical content courses. The technical courses in this model included the content and experiences one might expect to teach in a middle or high school program,

---

[7] Sander, Theodor: Structural aspects of teacher education in Germany today - a critical view. In:http://tntee.umu.se/publications/te_structure.html

as well as some additional content of greater technical sophistication. Until the late 20th century, virtually all of these technology teacher education faculty members had risen through the ranks of public (government) school teaching, so they had a very good understanding of what their students would encounter in the public schools upon graduation. They developed both the technical and pedagogy/professional education courses with that idea clearly in mind. Textbooks used in the technical courses were often the same texts used in high school courses, in part because they were helpful in preparing students for the future, and partly because the technology teacher education market was too small to support a different set of postsecondary technical books.

Split Faculty Model

Success of the technical, non-teaching degree options that were widely introduced in the 1970s led to a gradual, but significant change in the structure of technology teacher education in the majority of programs throughout the US. As the numbers of non-teaching (for example, Industrial Technology) students/majors quickly surpassed the numbers of Technology Education teaching students/majors, there was impetus to hire faculty for their industry experience and technical expertise rather than for their experience in the public school classrooms. By the 1990s, this new professoriate tailored their technical courses for the non-teaching majors, such as Industrial Technology, which typically outnumbered the teaching majors, sometimes by a very wide margin. Teacher education majors in this split faculty model enrol alongside the non-teaching majors in these technical courses, which are designed primarily to prepare students for industry. Typically, the equipment and processes taught are more sophisticated than would be appropriate for grades 6–12. These programs generally employ a small percentage of Technology Education faculty with public school teaching experience, who are responsible for teaching the pedagogy/professional education courses and supervising the clinical experiences in education. This split-faculty model now dominates the technology teacher education landscape in the US.

---

[8] http://www.comenius.de/projektedetail.cfm?id

Fifth Year Model

Post-secondary teacher education reform efforts in the 1980s promoted the idea of delivering professional education courses in a fifth year of post-secondary education to students who had earned baccalaureate degrees in the disciplines for which they were seeking teaching licensure. For example, a student might earn a baccalaureate degree in mathematics, and then enrol in a fifth year program that would lead to teaching licensure in mathematics. Under this model, technology teacher education programs would likely draw upon engineering graduates to enrol in a fifth-year licensure program, as engineering is the closest discipline to Technology Education. For a variety of reasons, including the fact that engineering graduates have generally been in great demand in the workplace, this model has not been widely implemented in technology teacher education.

Masters/Licensure Model

Many technology teacher education programs have long offered a masters/licensure option for students who already had baccalaureate degrees from various fields other than Technology Education. These masters/licensure programs generally require many of the undergraduate Technology Education courses to fulfill most of the licensure requirements, though some of the graduate courses may also 'double count' for licensure, lightening the course load somewhat. It has become common to structure these programs so students may earn a masters degree while concurrently fulfilling teacher licensure requirements, a process that can take two and a half years to complete. These masters/licensure options generally require at least two years for completion, depending upon the prior degree/course history brought in by each individual student.

Alternative Licensure Models

In the 1990s, critical teaching shortages in many different school subject areas led to the development of alternative teacher licensure models throughout the US. These are essentially shorter streamlined paths to teaching licensure implemented at the state level, and sometimes are administered separately from the post-secondary teacher education programs in the state. There are currently a wide variety of alternative technology teacher education models in use (Litowitz, 1998; Litowitz and Sanders, 1999). In virtually all cases, these alternative Technology

Education licensure options require a baccalaureate degree. Students in these options typically receive the professional education component and a relatively small number of technical content course hours. The student teaching experience is generally eliminated, and instead, a teacher/mentor is assigned to the first-year alternatively licensed teacher. In most states, those completing alternative licensure options must also pass exams that measure their professional education and discipline-specific knowledge, such as the Praxis I and II exams (Educational Testing Service, 2004). Typically, this alternative license is renewable within the state if additional education requirements are completed and upon successful completion of one or more years of teaching, though reciprocity agreements from one state to another may not apply to alternative licensure routes.

Trends in Technology Teacher Education

Standards for Technological Literacy

Throughout the US the *STL* (ITEA, 2000) have begun to facilitate change in technology teacher education programs. This change ranges from the subtle inclusion of new *STL* content in teacher education courses, to the complete restructuring of teacher education programs. *STL* certainly provides the impetus for an expansion of technological content and a rethinking of instructional method. Emphases on the study of design, the nature of technology, and the interaction between technology and culture, open the possibilities for reconceptualizing instructional activities. There is potential for greater emphasis on *knowing*, arguably with a corresponding decreased emphasis on *doing*, since perhaps a third of the standards focus upon cognitive understandings rather than on the tools and materials that have dominated the pedagogy of the profession over the past century.

Teacher Shortages

Significant teacher shortages have been a serious problem in the field for decades. In the 1970s, partly in response to the declining numbers of teacher education majors and partly in response to the substantial industry demand for their graduates in industry, many IA programs across the US began to offer non-teaching degree options, such as Industrial Technology. In general, their teacher education enrolments had been declining markedly, while their non-teaching option enrolments grew dramatically. Volk (1993) found graduates from these non-teaching options increased from 894 in 1970 to 7,063 in 1990. The movement to non-teaching options helps to explain the escalating Technology Education teacher shortages, which is now a grave problem in the profession (Householder, 1993; Litowitz, 1998; Vaglia, 1997; Volk, 1993, 1997, 2002; Weston, 1997). The total number of baccalaureate degrees granted to those preparing to teach in the field declined from 6,368 in 1970 (Volk, 2002) to an estimate of about only 800 in 2004. Hoepfl (1994) interviewed faculty from 20 discontinued technology teacher education programs, and found low enrolments to be a contributing factor in 19 of those 20 program closings. Householder (1993) identified 139 'operating' technology teacher education programs; compared to the finding of 70 active technology teacher education programs reported earlier in this chapter. Throughout the past decade, Technology Education has been formally designated a 'critical teaching shortage area' in states throughout the US. The longstanding nationwide shortage of licensed Technology Education teachers is a critical problem for the profession.

Based on the data reported earlier in this chapter, if the attrition rate (roughly the national average for all teachers in the US in 2004) is assumed as 8% among the estimated 36,000 Technology Education teachers currently employed in the US, there would have been a need for an estimated 2,880 newly licensed Technology Education teachers in fall 2004. This demand is approximately 3.5 times higher than the estimated 802 technology teacher education baccalaureate degrees awarded in 2004! If it was assumed that all of the 2004 Technology Education graduates accepted Technology Education teaching positions upon graduation, (a wildly optimistic assumption), the total number of 2004 technology teacher education graduates would have filled just over one fourth of the estimated 2880 Technology Education position openings that year. No national data are available for the percentage of current Technology Education teachers in the US who are hired temporarily without a license – often with little or no professional coursework in Technology Education – but these estimates suggest that figure is likely to be high.

The Diversity Dilemma

Technology teacher educators have been predominantly Caucasian males. The current number of female Technology teacher educators in the US may be counted on one hand and all minority populations are underrepresented in the technology teacher education faculty ranks. This is a

very serious problem for a field whose slogan is 'technological literacy for all.' The lack of diversity in technology teacher education has received considerable attention in the literature over the past two decades, including:
- the 1998 CTTE Yearbook, *Diversity in Technology Education* (Rider, 1998);
- publications resulting from the 'Women's [TE] Leadership Symposium' in 1996; and
- a number of journal articles addressing diversity (for example., Erekson and Trautman 1995; Liedtke, 1986; Liedtke, 1995; Markert, 2003; O'Riley, 1996; Rider, 1991; and Zuga, 1996).

In light of the current goals in the profession, the shortage of underrepresented populations in technology teacher education in the US should be deemed more serious today than ever before.

Engineering Education

The transition from IA to Technology Education in the 1980s was accompanied by growing interest in the 'design and technology' approach that evolved in the United Kingdom. By the end of the 20th century, the "technological problem solving method" (Savage and Sterry, 1990) had become a popular instructional approach in Technology Education programs across the US (Sanders, 2001). In the early 1990s, federal and state governments began to fund the development of Technology Education curriculum materials and in-service activities that facilitated integrated instruction in technology, science, and mathematics (see, for example, LaPorte and Sanders, 1995). Arguably, the use of the principles and processes of mathematics, science, and technology to solve technological problems is, in essence, 'engineering.' As early as 1992, the state of Virginia published course curriculum guides titled *Introduction to Engineering* and *Advanced Engineering*. In other words, Technology Education has been engaging with developmentally appropriate engineering content for more than a decade.

Few in the profession seemed to take notice when Bensen & Bensen (1993) suggested Technology Education embrace engineering content, nomenclature, and curricular organizers. But a great deal has changed since then. In the mid 1990s, the National Science Foundation (NSF) began funding engineering associations and faculty to develop educational materials and initiatives. In 1997, Project Lead the Way (PLTW) partnered with the College of Engineering at Rochester Institute of Technology to develop what is now a series of seven middle/high school pre-engineering courses. Currently that initiative utilizes about 20 university engineering programs to certify PLTW teachers who, in turn, offer PLTW courses in about 1,000 schools across the US (Blais, 2004). Massachusetts has developed integrated science, engineering and technology standards, resulting in the *Massachusetts Science and Technology/Engineering Curriculum Framework* (2001). Since 2002, the NSF has supported about 50 'Bridges for Engineering/Education' projects that brought the engineering and education faculties together at universities throughout the US. The National Academy of Engineering aggressively promoted a K–12 engineering education agenda in *Technically Speaking: Why All Americans Should Know More About Technology* (Pearson and Young, 2002). In 2003, the American Society for Engineering Education (ASEE) began a 'K–12 and Pre-College Division' to promote K–12 engineering education activities. In fall 2004, PLTW invited 15 teacher education programs to partner with them and begin to provide pre-service PLTW certification. Also in 2004, the NSF funded a *National Center for Engineering and Technology Education*, with goals that included increasing the professoriate and recruiting under-represented populations to Technology Education.

In the midst of this unprecedented flurry of activity from the engineering and Technology Education communities, 'engineering education' surfaced in 2004 as the hottest topic in TTE, as evidenced, for example, by the discussions that began to appear on the ITEA and CTTE Listservs and the Fall 2004 agendas at the Mississippi Valley and Southeast Technology Education Conferences. The role of engineering content in Technology Education and technology teacher education is likely to remain a principal educational issue in the years ahead.

AN EXAMPLE: THE COLLEGE OF NEW JERSEY

The technology teacher education program at the College of New Jersey (TCNJ) is a good example of a technology teacher education program that has moved forward with its curriculum. TCNJ is a medium-sized program situated in a state-supported institution with a history of preparing Industrial Arts/Technology Education teachers. The structure of the TCNJ's technology teacher education program is consistent with the

Traditional Four-Year Model described earlier. The College's technology teacher education program sits in the Department of Technological Studies, which is housed in the School of Engineering. The program currently has five full-time and two affiliated faculty staff who are teaching 115 technology teacher education majors: 60 are enroled in Technology Education and 55 in the new Math/Science/Technology – Elementary and Early Childhood Education option. Students in this option specialize in one of the five disciplines and may earn licensure in Technology Education through the middle school level.

TCNJ's technology teacher education program was arguably the first in the US to feature a design and technology approach. The 'Center for Design and Technology', established there in the early 1990s, provided a home for their two affiliated faculties and *TIES Magazine*, one of the goals of which was "to foster design-based problem-solving" (Anderson, 1988). This Center received funding for several large curriculum development grants that featured the D&T approach, including *Project UpDATE* (Todd, 1997) and *Children Designing and Engineering* (Hutchinson, 2002). Two of their faculty co-authored *Design and Problem Solving in Technology* (Hutchinson & Karsnitz, 1994), arguably the first Technology Education text in the US emphasizing a D&T approach. Currently, they are working to enhance the pre-engineering focus of their program as well (Karsnitz, personal communication, October 1, 2004). Their attention to design instruction (emphasized in STL) and new engineering content is consistent with what appear to be two important new directions for technology teacher education in the US.

TCNJ's technology teacher education program, like many in the US, includes a core of general education courses, pedagogy/professional education classes and a broad range of technical courses consistent with the aforementioned trends, current professional literature, the *STL,* and technology teacher education accreditation standards. The four-year course sequence is shown in Table 1.

Technology Teacher Education in the United States

Table 1.
Department of Technological Studies, 4-Year Course Sequence.

Fall		Spring	
Year 1			
Course	Component	Course	Component
First Year Seminar	Gen Ed	Academic Writing	Gen Ed
TST 161 Creative Design	Technical	LL Math/Science	Gen Ed
TST 171 Fundamentals of Technology	Technical	TST 111 Engineering Graphics	Technical
LL Math/Science	Gen Ed	TST 181 Structures & Mechanisms	Technical
		TST 191 Materials Laboratory	Technical

Fall		Spring	
Year 2			
Course	Component	Course	Component
LL SPE 203 Psy. Dev. child/Adol.	Gen Ed	LL Arts/Humanities	Gen Ed
TED 280 Introduction to Teaching	Prof Ed	LL Math/Science	Gen Ed
TST 281 Designing with Materials	Technical	TST 261 2D-Design	Technical
TST 231 Electronic Control	Technical	TST 291 Control Laboratory	Technical

Fall		Spring	
Year 3			
Course	Component	Course	Component
LL History (Technology in US)	Gen Ed	M/S Elective	Gen Ed
SPE 322 Inclusive Practices	Prof Ed	TED 380 JPE (280)	Prof Ed
TST 341 Biotechnical Systems	Technical	TED 460 Integrated MST for Learners	Prof Ed
TST 351 Robotics	Technical	TST 361 3-D Design	Technical
		TST 381 Prototyping Laboratory	Technical

Fall		Spring	
Year 4			
Course	Component	Course	Component
TED 480 Content & Methods	PE	LL Arts/ Humanities	Gen Ed
TED 481 Seminar	Prof Ed	TED 492 Facilities Design & Mgmt.	Technical
TED 490 Student Teaching	Prof Ed	TST 431 Designing Production Sys.	Technical
		TST 495 Senior Design	Technical

TEACHER CERTIFICATION

Teacher appointments were a local matter in the US until the middle of the 19th century, when states began to develop normal schools (teacher education programs) and provide funding for public elementary and secondary education. The need for greater accountability of state spending on education led to the development of State Boards of Education. Since the 1920s, the conventional way to earn a teaching license[3] in the United States was by graduating from a 'state approved program' (Darling-Hammond, 1999, p. 238) that met criteria established and monitored by state boards of education. Students graduating from such programs are thus assured of meeting the course and clinical experience requirements established by the state for initial teaching licensure. Currently, nearly all states utilize the state-approved program model. In addition, most states require teacher licensure candidates to pass one or more state or nationally developed/administered exams that assess professional and subject area knowledge.

Historically, state boards of education have regularly reviewed technology teacher education programs (for example, every 5 years) in order to sanction their 'approved program' status. Licensed teachers are required to take coursework and/or participate in professional in-service experiences in order to maintain their teaching license. States enter into 'reciprocity agreements' that allow teachers to be licensed immediately or with relative ease when moving from the state in which they earned their license to another state. For example, Virginia currently has reciprocity agreements with 48 other states.

Until the 1980s, teacher licensure regulations had been formulated independently by each of the 50 states, resulting in substantial variations from one state to another. Educational reform reports such as *A Nation Prepared: Teachers for the 21st Century* (Carnegie Forum on Education and the Economy, 1986) led to a general consensus of what all teachers should know and be able to do (Yinger, 1999). The Interstate Teacher Assessment and Support Consortium (INTASC) was formed to encourage cooperation/collaboration among states interested in rethinking teacher licensure standards. In the 1990s, the National Council for Accreditation of Teacher Education (NCATE) increased efforts to partner with states and professional associations in developing teacher education accreditation standards.

---

[3]Some states refer to these as teaching "certificates."

By the mid-1990s, there was a "remarkable consensus" on the ideals and standards for teacher licensure, accreditation, and certification (Yinger, 1999, p. 98). By the end of the century, NCATE had established partnerships with 45 states and the District of Columbia to "conduct joint reviews of colleges of education" (NCATE, 2000). In 1986, the ITEA and its affiliated Council on Technology Teacher Education (CTTE) voted to become an NCATE 'Specialized Professional Association,' on the speculation that NCATE affiliation would enhance the stature of the profession and assist the field in transitioning from IA to TE. Accordingly, the CTTE Accreditation Committee drafted the first ITEA/CTTE/NCATE technology teacher education Accreditation Standards, which took effect in 1987. The committee prepares revisions to these guidelines every five years, which are reviewed and approved by NCATE before going into effect. Significant changes were made in the 2004 revision of these accreditation standards to align them with STL. To date, about half of the active technology teacher education programs in the US have been through the NCATE accreditation process (CTTE, 2004). Through this evolving accreditation process, technology teacher education programs are probably becoming more alike from state to state than ever before, though substantial differences still remain.

CONCLUSION

As with most educational decision-making in America, the future of technology teacher education rests individually with the 50 US states. There are a number of critical issues and trends impacting technology teacher education in the US. Perhaps the most pressing is the pattern of declining enrolment that has plagued the field over the past three decades. This alarming technology teacher education enrolment decline – during a time when the children of the baby boom generation have been navigating post-secondary education in unprecedented numbers – has led to the downsizing of technology teacher education faculties and programs, a transition from the traditional technology teacher education model to the split-faculty technology teacher education model that requires fewer Technology teacher educators, and new 'alternative' pathways to teaching licensure. The resulting Technology Education teacher shortages have led to 'emergency hiring' in secondary Technology Education programs, in which unlicensed personnel are employed temporarily for up to three years. In many cases, secondary schools have downsized or closed their Technology Education programs. Over the past two decades, these practices have weakened the secondary and post-secondary Technology Education infrastructure in very significant ways.

While state licensure boards have approved alternative licensure pathways, standards for post-secondary teacher education have, paradoxically, become increasingly rigorous over the past two decades. Many believe this practice is undermining post-secondary teacher education, and ultimately, the overall quality of elementary and secondary education, as states and local divisions hire what some believe to be less qualified individuals to offset critical teacher shortages. The federal government has countered with the 'No Child Left Behind Act,' in an attempt to legislate 'highly-qualified teachers' for public school classrooms. The NCLB Act threatens elective subjects such as TE, as it attempts to mandate new performance standards in the 'academic' subject areas, which in turn will likely draw already limited resources away from Technology Education and other elective subjects.

Another trend is the serious decline in the number of doctoral granting institutions in the field from roughly a dozen such programs a decade ago to about half that number today. Teacher education has been under fire in the land-grant/research-centric institutions that once provided a steady stream of doctoral graduates to the profession. The now prevalent split-faculty technology teacher education model requires only about one third as many technology teacher education faculty staff as did the once-ubiquitous traditional technology teacher education model, resulting in a far smaller number of active Technology teacher educators in the US now compared with just several decades ago. In other words, the technology teacher education infrastructure has been vastly eroded over the past quarter century.

Meanwhile, Technology teacher educators across the US are dutifully working to address the curricular shifts motivated by the *STL* (ITEA, 2000), which are reflected in the recently updated ITEA/CTTE/NCATE accreditation standards. The very recent and considerable interest in increasing engineering content in the K–12 curriculum from both the Technology Education and engineering communities has caused technology teacher education to begin moving in that direction.

With the status of teacher education determined individually by each of the 50 states, technology teacher education is subject to influence from all of the aforementioned trends in different ways and degrees across the US. The politics of teacher licensure, teacher education program accreditation, teacher shortages, state and national standards, educational accountability, technological literacy, and the nature of alliances with potential allies, including engineering and science education, will conspire to shape the future of technology teacher education in the US in the decades ahead.

REFERENCES

Anderson, S. (1988, October). Letter to technology educators. *TIES Magazine.*

Battle, E. D. (1899). Manual training as related to girls. *Dissertation Abstracts International, 146*(67). (UMI No. AAG7233470)

Bensen, M. J. & Bensen, T. (1993). Gaining support for the study of Technology. *The Technology Teacher, 52*(6), 3–5, 21.

Blais, R. (2004, September). Presentation at the Project Lead the Way Meeting for Selected Technology Teacher Educators. Albany, NY, September 7, 2004.

Bonser, F. G. & Mossman, L. C. (1923). *Industrial arts for elementary schools.* New York: Macmillan.

Carnegie Forum on Education and the Economy. (1986). *A nation prepared: Teachers for the 21st century.* New York: Carnegie Council.

Council on Technology Teacher Education (2004). *ITEA/CTTE/NCATE accreditation standards.* Available: http://TechEd.vt.edu/CTTE/html/NCATE1.html.

Custer R. L. & Wright, R. T. (Eds.). (2002). Restructuring the Technology Teacher Education curriculum. in Ritz, Dugger, & Israel CTTE Yearbook #51. *Standards for Technological Literacy: The role of teacher education.* Peoria, IL: Glencoe/McGraw-Hill.

Darling-Hammond, L. (1999). Educating teachers for the next century: Rethinking practice & policy. In Griffin, G. A. (Ed.). *The education of teachers: Ninety-eighth yearbook of the National Society for the Study of Education.* Chicago: University of Chicago Press. pp. 221–256.

Dugger, W. E., Miller, C. D., Bame, E. A. Pinder, C. A., Giles, M. B., Young, L. H., & Dixon, J. D. (1980). *Report of the Survey Data.* Blacksburg, VA: Virginia Polytechnic Institute and State University.

Educational Testing Service (2004). The Praxis series: Professional assessments for beginning teachers. Available http://www.ets.org/praxis/

Erekson, T. L. & Trautman, D. K. (1995). Diversity or conformity? *Journal of Industrial Teacher Education, 32*(4), 32–42.

Gerbracht, C. & Babcock, R. (1969). *Elementary school industrial arts.* New York: Bruce.

Hoepfl, M. C. (1994). Closure of technology teacher education programs: Factors influencing discontinuance decisions. Doctoral dissertation, West Virginia University, 1994). *Dissertation Abstracts International*, 55–06A, 1535.

Householder, D. L. (1993). Technology teacher education: Status and prospect. *Journal of Technology Studies XIX* (1), 14–19.

Hutchinson, J. & Karsnitz, J. R. (1994). *Design and problem solving in technology education*. Albany, NY: Delmar Publishers, Inc.

Hutchinson, P. (2002). Children Designing & Engineering: Contextual learning units in primary design and technology. *Journal of Industrial Teacher Education, 39*(3), 122–145.

Industrial Arts Teacher Education at Oswego to 1941. (2004). Available: http://www.oswego.edu/tech/history.html No Author.

International Technology Education Association. (1996). *Technology for All Americans: A Rationale and Structure for the Study of Technology*. Reston, VA: Author.

International Technology Education Association. (2000). *Standards for technological literacy: Content for the study of technology*. Reston, VA: Author.

Karsnitz, J. (2004). Personal communication, October 1, 2004.

LaPorte, J. & Sanders, M. (1995). Technology, science, mathematics integration. In E. Martin (Ed.), *Foundations of technology education: Yearbook #44 of the council on technology teacher education*. Peoria, IL: Glencoe/McGraw-Hill.

Lewis, T. (1996). Accommodating border crossings. *Journal of Industrial Teacher Education 33*(2), 7–28.

Liedtke, J. (1986). Mentors and role models: Influences on the professional career. *The Journal of Epsilon Pi Tau 12*(1), 41–44.

Liedtke, J. (1995) Changing the organizational culture of technology education to attract minorities and women. *The Technology Teacher, 51*(6), 9–14.

Litowitz, L. S. & Sanders, M. E. (1999). *Alternative licensure models for technology education: Monograph #16 of the Council on Technology Teacher Education*. Reston, VA: Council on Technology Teacher Education.

Litowitz, L. S. (1998). Technology ed. teacher demand and alternative route licensure. *The Technology Teacher* 57(5), 23–28.

Markert, L. R. (2003). And the beat goes on: Diversity reconsidered. G. Martin & H. Middleton (Eds.). *Initiatives in technology education: Comparative perspectives.* Technical Foundation of America and the Centre for Technology Education Research, Griffith University.

Miller, W. R. & Boyd, G. (1970). *Teaching elementary industrial arts.* South Holland, IL: Goodheart-Willcox Company, Inc.

O'Riley, P. (1996). A different storytelling of technology education curriculum re-visions: A storytelling of difference. *Journal of Technology Education, 7*(2), 28–40.

Pearson, G. & Young, A. T. (2002). *Technically speaking.* Washington, DC: National Academy Press.

Rider, B. L. (1991). Problems and issues facing women in technology education. Paper presented at the Mississippi Valley Industrial Teacher Education Conference, Nashville, TN.

Rider, B. L. (Ed). (1998). *Diversity in Technology Education.* CTTE Yearbook #47. Peoria, IL: Glencoe/McGraw-Hill.

Sanders, M. E. (2001). New paradigm or old wine: The status of technology education practice in the US. *Journal of Technology Education, 12*(2), 35–55.

Sanders, M. E. (2003). The perplexing relationship between Technology education and career & technical education in the US. In G. Martin & H. Middleton (Eds.), *Initiatives in Technology Education: Comparative Perspectives.* San Marcos, TX: Technical Foundation of America.

Savage, E. & Sterry, L. (Eds.). (1990). *A conceptual framework for technology education.* Reston, VA: International Technology Education Association.

Schmidt, K. and Custer, R. L. (Eds.) (2004). *Industrial teacher education directory.* Reston, VA: Council on Technology Teacher Education.

Schmitt, M. L. & Pelley, A. L. (1966). Industrial arts education: A survey of programs, teachers, students, and curriculum. U. S. Department of Health, Education, and Welfare,. OE 33038, Circular No. 791. Washington, DC: Office of U.S. Government Printing Office.

Steeb, R. V. (1976, March). Funding industrial arts programs at the state level. *Man/Society/Technology,* 171–172.
Todd, R. (1997). A new paradigm for schooling. In J.J. Kirkwood & P. N. Foster (Eds.), *Elementary school technology education.* New York: Glencoe/McGraw Hill.
Warner, W. E., Gary, J.E., Gerbracht, C. J., Gilbert, H. G., Lisack, J. P. Kleintjes, P. L., & Phillips, K. (1947, April). *A Curriculum to Reflect Technology.* Paper presented at the annual conference of the American Industrial Arts Association.
Vaglia, J. (1997, October). *Technology education majors in colleges of the southeast: An enrollment overview.* Paper presented at the annual conference of the Southeast Technology Education Association, Roanoke, Virginia.
Volk, K. S. (1993). Enrollment trends in industrial arts/technology teacher education from 1970–1990. *Journal of Technology Education, (4)2,* 46–59.
Volk, K. S. (1997). Going, going, gone? Recent trends in technology teacher education programs. *Journal of Technology Education, (8)2,* 67–71.
Volk, K. S. (2002). *Enrollment trends in technology teacher education.* Paper presented at the Annual Conference of the International Technology Education Association, 3/15/02, Columbus, OH.
Weston, S. (1997). Teacher Shortage: Supply and demand. *The Technology Teacher, 57*(1), 6–9.
Winslow, L. L. (1922). *Elementary industrial arts.* NY: The Macmillan Company.
Yinger, R. J. (1999). The role of standards in teacher education. In Griffin, G. A. (Ed.). *The education of teachers: Ninety-eighth yearbook of the National Society for the Study of Education.* Chicago: University of Chicago Press.
Zuga, K. F. (1996, October). Women's ways of knowing and technology education. Paper presented at the Women's Leadership Symposium, Chicago.
Zuga, K. F. (1997, March 22). Technology education teacher education. Paper presented at the International Technology Education Conference in Tampa, FL.

Technology Teacher Education Summary

Chapter 13

P. John Williams
Edith Cowan University, Australia

INTRODUCTION

This book consists of discourses by experts in twelve countries of the technology education and technology teacher education systems in their countries. Some countries, which have quite homogenous educational systems and less diversity, such as New Zealand, make it easier to summarize and generalize about education. Other countries which have a greater degree of diversity make it very difficult to do so. Of course the bigger international picture is very diverse, and the danger in generalizing from too much diversity is that the dialogue becomes meaningless. However, among the twelve counties that have been covered, there are some common threads, and it is these that I would like to draw together.

Unlike school technology education, no international research has been conducted into the philosophical categorization of technology teacher education. A framework based on the rationale for the existence of a specific form of technology education was proposed by McCormick in 1993, and subsequently used by Black (1996) and Banks (1996). Four different types of justifications for technology education were identified.

1. The personal development opportunities it provides for students, for example practice in the solution of real problems and the types of thought processes associated with solving problems, and the multi-disciplinary approach to knowledge and information essential to technology education. This is stated as a rationale for technology education in Australia.

2. Education for the technological culture in which we live, to enable students to become informed decision makers and responsible users of technology, not so much for their own sake but for the benefit of society, is a significant rationale for technology education in South Africa.

3. The vocational dimension of technology education is a rationale that comes and goes with the passage of time, and tends to correlate with periods of national economic depression when policy makers and industrialists turn to education as part of the solution (Williams, 1998), for example the promotion of 'Workplace Knowhow' in schools in the USA, 'Essential Skills' in New Zealand, and 'Key Competencies' in Australia. The Technical and Vocational Education Initiative (TVEI) promoted technology education in the UK in the 1980's funded by the government department for industry, and technology education in the USA has often been identified with vocational education programs as a way to enhance funding opportunities.
4. Technology education as education for production was a strong rationale in many Marxist driven economies. With the collapse of the Soviet Union this rationale is less common, but was a driving force in Eastern European countries such as Hungary, Czechoslovakia, and the former German Democratic Republic, and Southern African countries such as Zimbabwe and Mozambique.

Generally speaking, the rationale for technology teacher education is essentially sympathetic to the system for which the teachers are being trained. The extent to which the rationale is different is reflected in a dissonance between the philosophy of the graduating teachers and the schools in which they are employed. This is also problematic during the periods of practice teaching, when experienced teachers are assessing trainees according to criteria that may be different from the training institution's philosophy.

In countries such as England where there is a very transparent competency base to teacher education, or countries such as Japan where the respective rationales are well established, such dissonance occurs infrequently. In some states of Australia and the USA, however, where the role of the university through teacher education is partly that of change agent, difficulties develop as the philosophy of technology education in schools is different to that in which teacher trainees are grounded.

An alternative categorization of approaches of technology education was proposed by DeVries (1994) which has been cited by many authors including Layton (1994), and Black (1996). The proposal included seven categories, which, while intended to relate to the school curriculum, could equally apply to technology teacher education programs.

1. A tradition in many countries has been the teaching of craft skills, often through the construction of set projects and the repetitive practice of relevant skills. This is a part of the basis of the Swedish tradition of Sloyd which has influenced the development of technical education in many countries throughout the world, including Russia.
2. In some instances technology education is organized along the lines of mass production, often in a business-like organizational framework. Relevant skills relate to the use of jigs and fixtures, a production line sequence of activity, control and organization. This is utilized in some Eastern European countries and to a lesser extent in a manufacturing technology context in the USA.
3. Though generally rejected as appropriate for technology, it may be organized as applied science, where technology is used in the teaching and learning of science. This approach is used in Denmark and Israel, where technology education in primary school and junior high school is integrated with science, and may not be dealt with as a valid end in itself.
4. A focus on technology as exclusively high or modern technology which is futuristic and emphasizes Information Technology. For example France, some programs in Israel and some of the learning modules available in the USA are inclined toward this approach.
5. Design, while a methodology of technology, may also be its organizational focus. In this case specific content is not so important, but rather the process through which students proceed in designing solutions to problems. Both the UK national syllabi and a number of approaches in Australia have been criticized for this approach, and the new curriculum in South Africa is inclined in this direction.
6. Technology may be structured as a series of problems to be solved, requiring information which is multi-disciplinary in nature. This is a common approach in parts of the USA.
7. The organization of content around the achievement of competencies is becoming a more common approach, evidenced for example in the UK, the 'competencies' in Australia and 'performance targets' in Sweden.

A more obvious and pragmatic (rather than philosophical) categorization of teacher education programs relates to the external pressures that are applied and reacted to in a very obvious and formative way. Any freedom to plan and act within the traditional university liberal arts continuum is more or less constrained by a range of pressures. Some examples follow.

1. **School Curriculum.** This pressure is an undeniably valid constraint as it forms the basis of the teachers' future professional activity. The school curriculum, and consequently its basis and approach is then either vicariously adopted by the teacher education program, or is accommodated within its broader philosophy. Countries in which a national school curriculum is implemented include England and Wales, New Zealand, France and Russia. The implication for these countries is that the content of teacher training is relatively similar, but there is still significant diversity in the approaches taken to train teachers.

2. **Teacher Shortage.** Many countries are experiencing a shortage of technology teachers. Despite the logic guaranteeing that employment should be a useful marketing tool to attract more students, this does not seem to be the case, with recruitment into technology teacher education remaining difficult. Smaller classes that result from declining enrolments can affect the viability of courses from a university management perspective. Economies of scale generated through large lecture groups are not consistent with a practical workshop approach to technology teacher education. Other impacts include the downsizing of technology teacher education departments and the closing of some departments, for example in Germany and the USA. The overall result is a general weakening of the technology education profession.

3. **Links for Survival.** In some contexts, generally where technology is not an element of the core school curriculum, links have to be made with other subject areas in order to ensure the continued viability of technology education. This has happened in some parts of the USA, for example where links have been made with mathematics and science, and more recently engineering.

4. **Legislation.** This can have a significant determining influence on the nature of technology teacher education programs. In Europe, the Bologna Process agreement for standardization of university programs across the European Union is having real impacts on teacher education programs, as seen in the cases in this book, for example in Germany and Russia. In England, legislation detailing the competencies for entry into teaching is having a very deterministic impact on both the structure and content of technology teacher education, and in the US, standards have become increasingly rigorous.

5. **Funding.** The recent change in status of universities in Japan to independent corporations is indicative of a new commercialization in universities in many countries. There is typically more competition between universities for students and research funding, and emphasis on the development of income generating programs. These pressures, in Japan, Hong Kong and Australia for example, have resulted in institutional integration. The resulting impact on technology teacher education programs has been a focus on economics. Given generally lower numbers of students and staff than other university faculties and higher costs associated with equipment, materials and workshops, a pure economics/business model does not sit well with technology teacher education programs. This has resulted in the closures of some programs and at least the need for continual justification of remaining programs.

6. **Emphasis on Basics.** The 'No Child Left Behind Act' in the USA and the national testing program in literacy and numeracy in Australia has the effect of threatening elective curriculum subjects such as technology education. While this has little direct effect on teacher education, it does tend to marginalize technology education as a less important area of education.

7. **Split Faculty Model.** A trend in a number of countries (Hong Kong, USA, Australia, Israel) is for the teacher education program to span a number of faculties. This is most often a change from the situation where the entire course was within one faculty, often a Faculty of

Education. Other faculties involved may include engineering, industrial design or art and design; where the students receive some of the technology content knowledge. One outcome of this trend is less technology education faculty involvement, with less professionals active in this area overall. This of course impacts on research and general professional development.

The majority of technology teacher education programs continue to battle against the legacy of technology education as a non-academic area of study dealing with a narrow range of skills and materials in a vocational or prevocational context. There is evidence that the focus of teacher education is distinctly general, broad and considers the social and environmental implications of technology without losing the important legacy of thinking in a practical context.

REFERENCES

Banks, F (1996) *Approaches and Models in technology Teachers Education: An Overview.* Journal of Design and Technology Education 1 (3) pp 197–211.

Black, P. (1996) *Curricular approaches and models in technology education.* Paper presented at JISTEC'96, Technology education for a changing future: theory, policy and practice. Jerusalem, January 8–11, 1996.

De Vries, M. (1994) Teacher education for technology education in Galton, M and Moon, B (eds) *Handbook of teacher training in Europe: Issues and trends.* London: The Council of Europe and David Fulton Ltd.

Layton, D. (1994) *Technology's challenge to science education.* Milton Keynes: Open University Press.

McCormick, R (1993) Technology education in the UK. In R. McCormick et al (eds) *Teaching and learning technology.* London: Addison Wesley Pub Co. p15–27.

Williams, P.J. (1998) *The Confluence of the Goals of Technology Education and the Needs of Industry: An Australian Case Study with International Application.* International Journal of Technology and Design Education, 8, 1–13.

INDEX

1917 revolution, 167

A

ACE (TE)
   see Advanced Certificate in Education (Teacher Education)
ACSFA
   see Advisory Committee on Pupil Financial Assistance
Adler, J., 211
Advanced Certificate in Education (Teacher Education) (ACE [TE]), 202, 203, 204
Advanced Engineering, 253
Advisory Committee on Pupil Financial Assistance (ACSFA), 114, 126
AEC
   see Australian Education Council
African National Congress (ANC), 193, 194
Aichi University of Education, 136, 139, 140–143, 146
Alternative Licensure Models, 250–251
Amalberti, R., 59, 67
American Industrial Arts Association, 245
American Society for Engineering Education (ASEE), 254
A Nation Prepared: Teachers for the 21st Century, 257
ANC
   see African National Congress
Anderson, S., 261
apartheid, 189, 193
applied science, 267
Armon, U., 114, 126
ASEE
   see American Society for Engineering Education
aspirantura, 180
Assessment, Competence Model of, 230–231
assessment standards for technology, illus., 191
Assistant Teacher Program (ATP), 18
Association of German Engineers, 72
ASTEC
   see Australian Science and Technology Education Council

ATP
   see Assistant Teacher Program
ATSE
   see Australian Academy of Technological Sciences and Engineering
attainment targets, 220–221
Australia, 265, 266, 267, 269
   certification of teachers in, 18–19
   degree programs in, 8–10, 12, 13–18
   technology teacher education in, 1–22
   vocational education in, 1, 5, 6
Australian Academy of Technological Sciences and Engineering (ATSE), 9
Australian Council for Computers in Education, 10
Australian Council for Education through Technology, 10
Australian Education Council (AEC), 1, 6, 7, 21
Australian Science and Technology Education Council (ASTEC), 3, 21
Avalos, B., 30, 42

B

Babcock, R., 244, 261
Backing Australia's Ability, 16
Bame, E. A., 245, 261
Banks, F., 118, 126, 215, 265, 271
Barak, M., 111, 114, 122, 126
Barlex, D., 118, 126
Barlow, J., 8–9, 11, 21
Barrow, H. S., 37, 42
Barzilai, A., 121, 122, 127
basics, emphasis on, 269
Battle, E. D., 244, 261
Ben-Gurion University of the Negev (BGU), 120, 126
Bensen, M. J., 253, 261
Bensen, T., 253, 261
Bessot A., 59, 67
BGU
   see Ben-Gurion University of the Negev
Black, P., 265, 266, 271
Blais, R., 261
Bologna Process, 72–73, 77, 80

Index

Bonser, F. G., 244, 261
Boyd, G., 244, 263
Bridges for Engineering/Education project, 254
British Design and Technology model, 193
Brown, H., 94, 108
Brunner, J. J., 27, 42
Bundesministerium für Bildung und Forschung, 87
Bundesregierung, 87
Burns, J., 147, 165
Bushbuckridge Master Programme for Technology Teachers, 206
Business Educators of Australia, 11

C

C&D
  see Craft and Design
CAD/CAM software, 222, 223, 227
California University of Pennsylvania, 246
Career and Technical Education (CTE), 242
Carnegie Forum on Education and the Economy, 257
Carr, M., 152, 156, 165
CATTS
  see Center to Advance the Teaching of Technology and Science
Cazenaud, M., 59, 67
CCEA
  see Common Curriculum Examination Authority
CDT
  see Craft, Design and Technology
Center for Design and Technology Education, 232
Center for Science and Technology Education Research, 152
Center to Advance the Teaching of Technology and Science (CATTS), 242
Centre d'Étude et de Recherche sur l'Emploi et les Qualifications (CEREQ), 46
CEREQ
  see Centre d'Étude et de Recherche sur l'Emploi et les Qualifications
certification of teachers
  in Australia, 18–19
  in Chile, 31, 36, 37–39
  in France, 63–65
  in Germany, 86
  in Hong Kong, 106–107
  in Israel, 123–124
  in Japan, 143–144
  in New Zealand, 162–163
  in Russia, 186
  in South Africa, 208–209
  in the United Kingdom, 235–237
  in the United States, 257–258
Challenge 2000, 62, 65
Charlot, B., 47, 67
Chester, I., 12, 21
Children Designing and Engineering, 255
Chile
  certification of teachers in, 31, 36, 37–39
  degree programs in, 31–33
  Ministry of Education in, 24, 26, 29, 33, 34, 39, 40, 43
  technology education in, problems and progress, 24–26
  technology teacher education in, 23–43
  vocational education in, 23, 24, 29, 31
Chile Califica (Chile Qualifies), 26, 36
Chilean Educational Reform, 39
Chinese University, 94
Chisholm, L., 211
Christie, P., 211
CNP
  see Conseil National des Programmes
Coles, A. S., 114, 126
College of Education, 94
College of New Jersey, The, 254–255
Committee on Higher Education, 208
Common Curriculum Examination Authority (CCEA), 222, 238
Compton, V., 152, 165
Conceptual Framework for Technology Education, 245
concurrent model, 226
Conference of the Ministers of Education and Culture, 79
consecutive model, 226
Conseil National des Programmes (CNP), 46

273

Index

Constitutional Teaching Law (LOCE), 27, 39
core competences for trainees, 224–225
Council of Australasian Media Education Organizations, 10
Council on Technology Teacher Education (CTTE), 253, 254, 258, 259
Cowie, B., 165
Cox, D., 35, 42
Craft and Design (C&D), 220, 223
Craft, Design and Technology (CDT), 220
craft skills, 267
CREQ
   see Criteria for the Recognition and Evaluation of Qualifications for Employment in Education
Criteria for the Recognition and Evaluation of Qualifications for Employment in Education, 198
CS
   see Curriculum Studies
CTE
   see Career and Technical Education
CTTE
   see Council on Technology Teacher Education
Culture of the House and Craft-applied Art Creativity, 183
curriculum, 268
Curriculum 2005, 193–194, 195, 199
Curriculum Cymreig, 215
Curriculum Development Committee, 90, 108
Curriculum Development Council, 89, 108
Curriculum Development Institute, 95
Curriculum Review Committee, 194
Curriculum Studies (CS), 207–208
Curriculum to Reflect Technology, A, 245
Custer, R. L., 246, 248, 261, 263
Czechoslovakia, 266

D

D&T
   see Design and Technology
Dakers, J. R., 223, 238
Darling-Hammond, L., 257, 261
DATA
   see Design and Technology Association
DEC
   see South Africa Department of Education and Culture
degree programs
   in Australia, 8–10, 12, 13–18
   in Chile, 31–33
   in France, 52–58
   in Germany, 77–80
   in Hong Kong, 94, 95, 97–98
   in Israel, 117–124
   in Japan, 136–140
   in New Zealand, 153, 157–160
   in Russia, 172, 175, 177–188, 180–181
   in South Africa, 196
   in the United Kingdom, 226–229, 232
   in the United States, 247–248
DENI
   see Department of Education Northern Ireland
Denmark, 267
DEP
   see Direction des Études et de la Prospective
Department for Education/Welsh Office, 239
Department of Education Northern Ireland (DENI), 216, 217, 238, 239
design, 267
Design and Make Assignments (DMAs), 221
Design and Problem Solving in Technology, 255
Design and Technology (D&T), 89–106, 217–218, 221, 222, 225, 226, 234
   Bachelor of Education in, 98–103
   Post Graduate Degree in, 103–106
   as a school subject, 92–93
design and technology approach, 253
Design and Technology Association (DATA), 218, 221, 222, 223, 225, 231, 238
Design and Technology, Center for, 255
Design and Technology: Systems and Control in England and Wales, 220
Design in Education Council Australia, 10
Develay, M., 47, 67

De Vries, J. M., 125, 127
De Vries, M., 266, 271
Dewey, John, 244
Direction des Études et de la Prospective (DEP), 46n
diversity, 252–253, 265
Diversity in Technology Education, 253
Dixon, J. D., 245, 261
DMAs
    see Design and Make Assignments
DNE
    see South Africa Department of National Education
Doherty, R., 223, 238
Domestic Science, 217–218
Doppelt, Y., 114, 126
Drucker, P. F., 111, 127
Dual Education, definition of, 26
Duffy, T., 37
Dugger, W. E., 245, 261

E

East Germany
    see German Democratic Republic
Eberhard, Kurt, 87
EC
    see European Commission
Edith Cowen University, 1, 15–18
Education and Manpower Branch, 108
Education and Manpower Bureau, 95, 106
Education Commission, 91, 92, 108
Education Labour Relations Council, 209
Education Orientation Law, 50, 53
Education Renewal Strategy (ERS), 192
Educational Personnel Certification Law, 135
EHEA
    see European Higher Education Area
Elton, F., 24, 25, 34, 42
Engineering Council, The, 221
engineering education, 253–254
Engineering, School of, 255
England, 266
Environmental Studies, 219–220
Erekson, T. L., 253, 261
ERS
    see Education Renewal Strategy

ESIB
    see National Unions of Students in Europe
Essential Skills, 266
EUA
    see European University Association
European Commission (EC), 73
European Higher Education Area (EHEA), 73–74
European University Association (EUA), 73
Exams, Praxis I and II, 251

F

Faculty of Technology and Enterprise, 174
Favier, J., 59, 67
Federal Component of the State Standards for General Education, 170, 187
FET
    see Further Education and Training Band
Fifth Year Model, 250
First Degree, 52–53
Five-Year Continuous Program, 177–179
Flensburg University, 80–86
    non-teacher training course of study, 85–86
    structure of studies, illus., 80
    teacher training course of study at, 81–85
Forsyth, A. J., 114, 127
France
    certification of teachers in, 63–65
    degree programs in, 52–58
    Ministry of Education in, 46, 47, 51, 56, 57, 64
    organization of curriculum in, 46–51
    school system of, 45–46
    technology education in, 46–51
    technology teacher education in, 45–67
    vocational education in, 45–50
Frank, M., 121, 122, 127
Fullan, M., 127
Fundamental Law of Education, 132
Furlong, A., 114, 127
Further Education and Training Band (FET), 189, 193, 194, 195, 196, 203, 209

Index

G

Gary, J. E., 245, 264
GC
  see Graphic Communication
General Education and Training Band (GET), 189, 193, 194, 195, 203, 207, 209
General Teaching Councils (GTC), 235–236
General Technology Department, 174
Gerbracht, C., 244, 261
Gerbracht, C. J., 245, 264
German Democratic Republic, 75, 266
Germany
  certification of teachers in, 86
  degree programs in, 77–80
  school system in, 69–70
  technology teacher education in, 69–87
GET
  see General Education and Training Band
Gibson, J., 8–9, 11, 21
Gijselaers, W., 37
Gilbert, H. G., 245, 264
Giles, M. B., 245, 261
Ginestié, J., 45, 48, 67
Giroux, H., 118, 127
Graduate Teacher Training Registry (England and Wales) (GTTR), 229
Graphic Communication (GC), 220, 223
GTC
  see General Teaching Councils
GTTR
  see Graduate Teacher Training Registry (England and Wales)
guidelines, subject area, 223–224
Gurevich, M., 187
Gysling, J., 35

H

Harber, C., 211
Harlow, A., 165
Hartshorne, K., 211
HEFCW
  see Higher Education Funding Council Wales
HEQF
  see Higher Education Quality Committee

Herzen State Pedagogical University, 173, 174, 181–186
HET
  see Higher Education and Training Band
Hidetoshi Miyakawa, 129
Higher Education and Training Band (HET), 189, 195, 199, illus., 200
Higher Education Funding Council Wales (HEFCW), 229
Higher Education Quality Committee (HEQF), 208
Hill, A-M., 48, 67
HKCAA
  see Hong Kong Council for Academic Accreditation
HKIEd
  see Hong Kong Institute of Education
HKSAR Chief Executive, 93, 94, 95, 109
Hochschul-Rektoren-Konferenz, 87
Hoepfl, M. C., 252, 262
Hoepken, G., 69, 72n, 87
Home Economics Institute of Australia, 10
Hong Kong, 269
  certification of teachers in, 106–107
  degree programs in, 94, 95, 97–98
  Education Department in, 92, 94, 95, 96, 97, 109
  schools, background of, 92
  technical schools in, 89–90
  Technical Teachers College, 95–96
  technology teacher education in, 89–109
Hong Kong Advanced Level Examination, 94
Hong Kong Council for Academic Accreditation (HKCAA), 98
Hong Kong Institute of Education (HKIEd), 94, 95, 96, 97, 107, 109
Hong Kong University, 94
Householder, D. L., 248, 252, 262
HSRC
  see Human Sciences Research Council
Human Sciences Research Council (HSRC), 192
Hungary, 266
Hutchinson, J., 262
Hutchinson, P., 255, 262

Index

I

IA
  see Industrial Arts
ICT, 235
Iglesias, J., 23, 37, 42, 43
imaginaries, 173
Industrial Arts (IA), 244, 245, 248, 251, 253
Industrial Arts and Homemaking Education, 129, 132–135
Industrial Arts Teacher Education at Oswego to 1941, 246
industrialisation, 174
industrial project method, 48, 57, 61, 62
Industrial Teacher Education Directory (ITE), 246, 247
Industrial Technology, 248, 249
Industry-Pedagogy faculty, 174
Information Technology, 2, 3, 267
INTASC
  see Interstate Teacher Assessment and Support Consortium
International Technology Education Association (ITEA), 23–24, 43, 72, 109, 242, 244, 245, 251, 254, 258, 259
Interstate Teacher Assessment and Support Consortium (INTASC), 257
Introduction to Engineering, 253
Israel, 267
  certification of teachers in, 123–124
  degree programs in, 117–124
  graduate studies in, 120–121
  in-service training in, 119–120
  pre-service training in, 118–119
  technology teacher education in, 111–127
  vocational education in, 114
Israel Institute of Technology, 121–123
  see Science-Technology-Society
IT
  see Information Technology (IT)
ITE
  see *Industrial Teacher Education Directory*, 254
ITEA
  see International Technology Education Association

IUFM
  see university teacher training institute
IUFM Uniméca, Aix-Marseille, 52, 59, 59–63
  general background of, 59–60
  course organization at, 60–63

J

Jansen, J., 211
Japan, 266, 269
  certification of teachers in, 143–144
  degree programs in, 136–140
  education in, and gender divisions, 129, 132, 133–134
  Ministry of Education, Science, Sports and Culture, 131, 142, 143, 145, 146
  technology teacher education in, 129–146
  vocational education in, 132–133
Japanese Society of Technology Education, 135
Jerusalem International Science and Technology Education Conference (JISTEC), 125
JISTEC
  see Jerusalem International Science and Technology Education Conference
Jones, A., 147, 151, 152, 155, 165
Jouineau, C., 67
justifications for technology education, 265–266
  masters courses in, 155
  Russian, aims of, 168
  in secondary schools, 4–5
  structure in United States, 243–244
  teacher shortage in, 11
  transition from industrial arts, 245

K

Kahn, M. J., 211
Kandidat Nauk, 172
Karsnitz, J. R., 255, 262
Kelson, A. C., 42
Key Competencies, 266
Key Learning Areas (KLA), 89
key stages, 216–217
King, K., 211
KLA
  see Key Learning Areas

Index

Kleintjes, P. L., 245, 264
KMK
   see Ministers for Education Conference
Know How 2, 157
Koeshunov, T. Y., 185, 186, 187
Korshunova, N. N., 185, 186, 187
Kraak, A., 192, 193, 211

L

LaPorte, J., 262
LAT
   see Limited Authority to Teach
Layton, D., 266, 271
Lazutova, M. N., 187
Learning-Teacher Methodology Union (UMO), 179
Lednev, V. S., 187
Lend a Hand, 47
Lewin, K., 211
Lewis, T., 245, 262
Li, Professor Arthur K.C., 95
Liedtke, J., 253, 262
Limited Authority to Teach (LAT), 19
Lisack, J. P., 245, 264
Litowitz, L. S., 250, 252, 262, 263
LOCE
   see Constitutional Teaching Law
Luthuli, D., 211
Lux, D., 59, 67

M

Maori, 148, 148n
Markert, L. R., 253, 263
Massey University, 156
Masters/Licensure Model, 250
Mather, V., 152, 165
Maths, Science and Technology (MST), 206
McCaslin, N. L., 124, 127
McCormick, R., 217, 239, 265, 271
MCEETYA
   see Ministerial Council on Education, Employment, Training and Youth Affairs
McKenzie, D., 147, 165
McKibbin, M., 124, 127
MCTE
   see Ministerial Committee on Teacher Education

Mendelson, N., 126
Metropolitan Technological University, 28
Mewe, Fritz, 87
Miller, C. D., 245, 261
Miller, W. R., 244, 263
Millersville University of Pennsylvania, 246
Ministerial Committee on Teacher Education (MCTE), 197, 201
Ministerial Council on Education, Employment, Training and Youth Affairs, 1, 11, 21
Ministerial Task Group Reviewing Science and Technology Education, 148
Ministers for Education Conference (KMK), 72
Ministers of the Crown, 148
Ministry of Education and Culture, 112, 114, 127
Ministry of Education of the Russian Federation, 176, 179, 180, 187
Mississippi Valley and Southeast Technology Education Conferences, 254
Modernisation, Strategy of, 169
Modular Curricular Approach, 26
Moreland, J., 151, 165
Morris, E., 227, 239
Morrow, M., 211
Mossman, L. C., 244, 261
Mothupi, C., 189
Mouton, J., 211
Mozambique, 266
MST
   see Maths, Science and Technology
Muñoz, J., 28, 43
Murray-Smith, S., 5, 21

N

National Academy of Engineering, 254
National Association of Agricultural Educators, 11
National Center for Engineering and Technology Education, 254
National Certificate in Educational Achievement (NCEA), 159
National Council for Accreditation of Teacher Education (NCATE), 257–258, 259

National Education Policy Act of 1996, 195, 197
National Education Policy Investigation, 211
National Facilitator Training Programme, 152
National Fund for Personal Training, 169
National Higher Education Council, 23
National Qualifications Framework (NQF), 193, 197, 198, 199, 208
National Science Foundation (NSF), 253, 254
National Task-Force for the Advancement of Israel's Education (NTAIE), 113, 118, 124, 127
National Unions of Students in Europe (ESIB), 73
NCATE
   see National Council for Accreditation of Teacher Education
NCC, 239
NCEA
   see National Certificate in Educational Achievement
NCLB
   see No Child Left Behind Act
New Jersey, The College of, 254–255
Newly Qualified Teacher, 236
New South Wales, 4
New Zealand, 265–266
   certification of teachers in, 162–163
   degree programs in, 153, 157–160
   graduate education in, 159–160
   Ministry of Education in, 148, 149, 152, 157, 162, 165
   Ministry of Research, Science and Technology in, 148, 165
   pre-service education in, 154–155
   technology teacher education in, 147–166
New Zealand Curriculum Framework, 148, 156
New Zealand Teachers' Council, 162
New Zealand Technology Curriculum, 158
   Teacher Support Services, 163
   technological knowledge, 149–150
   technological capability, 150–151
   technology and society, 151

NGO
   see non-governmental organizations
NHEC
   see National Higher Education Council
Nikandrov, N. D., 187
No Child Left Behind Act (NCLB), 259
non-governmental organizations, 195–196, 197, 203, 204
Norambuena, C., 28
Nordnflycht, M. E., 42
normal schools, 246n
Norman, E., 59, 67
Norms and Standards for Educators (NSE), 198
Northrhine-Westfalia Expert commission, 79
NQF
   see National Qualifications Framework
NQT
   see Newly Qualified Teacher
NSE
   see Norms and Standards for Educators
NSF
   see National Science Foundation
NTAIE
   see National Task-Force for the Advancement of Israel's Education
Núñez, I., 28, 43

O

OBE
   see Outcomes-based education
OECD
   see Organisation for Economic Co-operation and Development
Oliver, B., 124, 127
Organisation for Economic Co-operation and Development (OECD), 36, 43
   PISA study, 71–72, 77, 78
O'Riley, P., 253, 263
Ortega, L., 28
ORT-STEP Institute, 203
Osterkamp, S., 87
Outcomes, 189, 190–191
Outcomes-based education (OBE), 193, 194, 198, 199, 207

Index

P

Parker, B., 198, 211
Parks, D., 124
Pavlova, M., 167, 187
PBL
   see problem-based learning or project-based learning
Pearlman-Avnion, S., 122, 126
Pearson, G., 254, 263
Pelley, A. L., 245, 263
Pérez, R., 28
Perkins, D. N., 149, 165
Phillips, K., 245, 264
Pinder, C. A., 245, 261
PISA study, 71–72
Piskunov, A. I., 173, 187
Pitt, J., 187
PLTW
   see Project Lead the Way
Policy Studies (PS), 207–208
Post-Graduate Degree in Education for Design and Technology, 103–106
pressures on technology education, 268–269
PREST, 47
Preston, B., 11, 21
primary and middle school teacher education, 157–158
primary and secondary education, technology teacher education for, 32–33
primary schools, technology education in, Australia, 2–3
primary technology teacher education, 12–13
Prime Minister's Science, Engineering and Innovation Council, 3, 21, 22
problem-based learning (PBL), 30, 36–39
professional institutes, 23, 24, 27, 28, 29, 34, 39
programmes, modularization of, 227
Program of vocational education development, 171, 187
project-based learning (PBL), 119, 121–122, 125, 190–191
Project Lead the Way (PLTW), 253–254
Project UpDATE, 255
PS
   see Policy Studies

Q

QTS
   see Qualified Teacher Status
Qualified Teacher Status (QTS), 225, 232, 236–237

R

Rak, I., 59, 67
Ramsey, G., 22
Rationale and Structure for the Study of Technology, 245
Ray, W., 59, 67
Raz, E., 122, 126
Reed, Y., 211
Reich, G., 87
Review of vocational education development in 2001–2002, 171, 187
Rhodes University, 203–206
Rider, B. L., 253, 263
Robinson, M., 211
Robinson, P., 215, 221, 239
Rogan, J., 211
Russia, 273
   certification of teachers in, 186
   degree programs in, 172, 175, 177–178, 180–181
   education system in, 171–172
   modernization of higher education in, 175–176
   technology education in, and gender divisions, 168–169
   technology teacher education in, 167–188
   vocational education in, 172, 173, 187
Russian Education by 2001: Analytical Overview, 187
Russian Education: Federal Portal, 175, 187
Russian government, 187

S

SACE
   see South African Council of Educators
Salomon, G., 149

Index

Samuel, M., 211
Sander, Theodor, 76n, 248n, 87
Sanders, M., 241, 262, 263
Sanders, M. E., 245, 250, 262
Santiago, 28
SAQA
    see South African Qualifications Authority
Sasova, I., 187
Savage, E., 245, 253, 263
Savery, J. R., 37, 43
Sayed, Y., 211
SCCC, 219, 239
Schleswig-Holstein Technology Curriculum, 70
Schmayl, W., 75n, 87
Schmidt, K., 246, 263
Schmitt, M. L., 245, 263
School-Centered Initial Teacher Training (SCITT), 227
School Education Law, 132
Science-Technology-Society (STS), 112, 118–119
SCITT
    see School-Centered Initial Teacher Training
Scobey, 244
Scottish and Executive Education Department (SEED), 229
Scottish Credit and Qualifications Framework, 228
Scottish Office Education and Industry Department (SOEID), 231, 239
Secondary School Places Allocation (SSPA), 91
Secondary schools, technology education in, 4–5
secondary teacher education, 158–159
secondary technology teacher education, 13–15
secondary vocational schools, technology teacher education for, 35–36
SEED
    see Scottish and Executive Education Department

Sekhukhuni Master Science and Technology Education Programme, 206
Sheffield Hallam University, 232–235
shortage of teachers,
    in Australia, 11
    in general, 268
Shulman, L. S., 117, 127, 152, 165
Sills, N., 43
Sloyd, Swedish tradition of, 267
Smithers, A., 215, 221, 239
Smith-Hughes Vocational Education Act, 250–251
SOEID
    see Scottish Office Education and Industry Department
Soto, R. F., 43
Sound of the System learning unit, 122–123
South Africa, 265, 267
    certification of teachers in, 208–209
    degree programs in, 196
    education in, and racial divisions, 192, 195
    history of education in, 192–194
    Minister of Education in, 198, 208
    teacher education in, 189–190
    technology teacher education in, 189–213
    three pillars of education system in, 193
South Africa Department of Education, 189, 199, 209, 211
South Africa Department of Education and Culture (DEC), 211
South Africa Department of National Education (DNE), 192, 197, 211
South African Council of Educators (SACE), 208–209
South African Qualifications Act of 1995, 195
South African Qualifications Authority (SAQA),193, 197, 198, 208
South African Qualifications Authority Act of 1995, 197
South African School Act of 1996, 195
split faculty model, 249, 269
SSPA
    see Secondary School Places Allocation
St. Petersburg, 173, 181

281

Index

Standards for Technological Literacy (STL), 241, 242, 251, 255, 258, 259
State University of New York at Oswego, 246
Statement on Technology for Australian Schools, 1, 6
Statistics of Russian Education, 172, 187
Steeb, R. V., 245, 264
Sterry, L., 245, 253, 263
Stevens, A., 189, 211
STL
   see Standards for Technological Literacy
STL: Content for the Study of Technology, 245
Structure Plan of the German Council on Education, 76
STS
   see Science-Technology-Society
study, nature of, 230–231
Sweden, 267

T

T&D
   see Technology and Design
Tamir, A., 111, 125, 126, 127
Tapp, J., 211
Tasmania, 4
Tatischev, 173
TCNJ
   see College of New Jersey, The
TE
   see Technology Education
Teacher Training Agency (England) (TTA), 221, 227, 231, 236, 239
Technical and Vocational Education Initiative (TVEI), 220, 222, 266
Technic and Technical Creativity, 183
Technical Pedagogical Institute, 28
Technical State University (TSU)
   see University of Santiago
technical training centres, 23, 24, 27, 29
Technically Speaking: Why All Americans Should Know More About Technology, 254
technikons, 189
Technion
   see Israel Institute of Technology
Technological State University, 27
Technological Studies (TS), 220, 222–223
Technological Studies, Department of, 255

Technology and Children, 244
Technology and Design (T&D), 216, 222
Technology and Design Ministerial Report, 222
Technology as a non-academic area, 269
   Russian definition of, 168
Technology Design in Northern Ireland, 220
technology education (TE), 241–242
   alternative categorization of approaches, 266–267
   as elective in grades 6–12, 243
   in grades K–5, 243–244
Technology Education Collaborative Laboratory, 38
Technology Education Council for Children, 244
Technology education history
   in Australia, 5–7
   in Chile, 27–28
   in France, 49–51
   in Germany, 74–75
   in Hong Kong, 93–95
   in Israel, 114–116
   in Japan, 132–135
   in New Zealand, 152–153
   in Russia, 173–175
   in South Africa, 192–194
   in the United Kingdom, 220–223
   in the United States, 244–245
Technology Education Key Learning Area (TEKLA), 93
Technology in the New Zealand Curriculum, 149
Technology Studies (TS), 207
Technology Teacher Development Resource Package, 152
technology teacher education (TTE), at
   Aichi University of Education, Japan, 140–143
   College of New Jersey, United States, 254–255
   Edith Cowan University, Australia, 15–18
   Flensburg University, Germany, 80–86
   Herzen State Pedagogical University, 181–186
   Hong Kong Institute of Education, 97–106
   Israel Institute of Technology, 121–123

IUFM Aix-Marseille, France, 59–63
Rhodes University, South Africa, 203–206
Sheffield Hallam University, United Kingdom, 232–235
University of Atacama, Chile, 36–39
University of Limpopo and PROTEC, South Africa, 206–208
University of Waikato, New Zealand, 160–162
technology teacher education overview
 in Australia, 7–11
 in Chile, 29–31
 in France, 51–53
 in Germany, 76–77
 in Hong Kong, 95–97
 in Israel, 116–118
 in Japan, 135–136
 in New Zealand, 154–156
 in Russia, 175
 in South Africa, 194–197
 in the United Kingdom, 223–226
 in the United States, 246–247
technology teacher education structure
 in Australia, 11–15
 in Chile, 31–36
 in France, 54–59
 in Germany, 77–80
 in Hong Kong, 97–106
 in Israel, 118–121
 in Japan, 136–140
 in New Zealand, 156–160
 in Russia, 176–181
 in South Africa, 197–203
 in the United Kingdom, 226–231
 in the United States, 247–254
technology teacher education trends, 12, 251–254
Teixido, C., 59
TEKLA
 see Technology Education Key Learning Area
Thatcher government, 215, 220
Third International Mathematics and Science Study (TIMSS), 78, 112, 113, 127
TIES Magazine, 255

TIMSS
 see Third International Mathematics and Science Study
Todd, R., 255, 264
Tokyo Shoseki, 146
Trautman, D. K., 253, 261
TS
 see Technological Studies
tsarist government, 173
TSU
 see University of Santiago
TTA
 see Teacher Training Agency (England)
TTE
 see technology teacher education
Tung Chee Hwa, 93
TVEI
 see Technical and Vocational Education Initiative

U

UCAS
 see Universities and Colleges Admissions Service (UK)
UK
 see United Kingdom
UMO
 see Learning-Teacher Methodology Union
UMOa, 179, 187
UMOb, 187
UNESCO
 see United Nations Educational, Scientific and Cultural Organization
United Kingdom, 266, 267
 certification of teachers in, 235–237
 degree programs in, 226–229, 232
 Department for Education and Skills in, 218, 238
 Department of Employment in, 220
 Minister for Schools in, 227
 technology teacher education in, 215–239
United Nations Educational, Scientific and Cultural Organization (UNESCO), 171
United States, 266, 267, 269
 certification of teachers in, 257–258
 degree programs in, 247–248

Index

non-teaching degree options in, 249, 251–252
technology education structure in, 243–244
technology teacher education in, 241–264
vocational education in, 245
Universidad Austral de Chile, 28
Universidad Católica de Chile, 28
Universidad de Concepción, 28
Universidad Técna del Estado, 28
Universities and Colleges Admissions Service (UK) (UCAS), 229
University of Atacama, 28, 36–39
University of Chile, 27, 28
University of Hamburg, 79
University of Limpopo and Protec, 206–208
University of Santiago, 28
University of Science and Technology, 107
University of Waikato, 156, 160–162
University of Wisconsin-Stout, 246
university teacher training institute (IUFM), 51, 57, 58, 59, 64, 65
Up-bringing House, 173–174
upper secondary schools in France, 49, illus., 50

V

Vaglia, J., 252, 264
Vera, R., 42
Vérillon, P., 59, 67
Vocational and Homemaking Education Course, 132–133
vocational education, 266
 in Australia, 1, 5, 6
 in Chile, 23, 24, 29, 31
 in France, 45–50
 in Israel, 114
 in Japan, 132–133
 in Russia, 171–174
 in the United States, 245
Vocational Education Act, 245
Volk, K. S., 89, 92, 93, 109, 251–252, 264
Volmink, J. D., 211

W

WACOT
 see Western Australian College of Teaching
Wales, 5
Warner, W. E., 245, 264
Watts, D., 8, 9, 11, 22
Western Australia, 4
Western Australian College of Teaching (WACOT), 18
Weston, S., 252, 264
Wilkerson, L., 37, 43
Williams, A., 8, 12, 22
Williams, A. P., 5, 22
Williams, J., 8, 12, 22
Williams, P. J., 1, 2, 3, 8, 12, 22, 266, 271
Winslow, L. L., 244, 264
Women's [TE] Leadership Symposium, 253
Workplace Knowhow, 266
World War II, 245
Wright, R. T., 248, 261

Y

Yager, R. E., 112, 127
Yehiav, R., 126
Yinger, R. J., 257, 258, 264
Yip, W. M., 93, 109
Yon, R., 11, 22
Young, A. T., 254, 263
Young, L. H., 245, 261
Young, M., 211

Z

Zimbabwe, 266
Zuga, K. F., 248, 253, 264